HONORING OUR
DETROIT RIVER

➔ ←

HONORING OUR DETROIT RIVER

CARING FOR OUR HOME

EDITED BY

John H. Hartig

PREFACE BY

Congressman John D. Dingell

Library of Congress Cataloging-in-Publication Data

Honoring our Detroit River : caring for our home / edited by John H.
Hartig.

    p.   cm.

Includes bibliographical references (p.   ).

ISBN 0-87737-044-3 (cloth : alk. paper)

1. Stream ecology—Detroit River (Mich. and Ont.)   2. Detroit River
(Mich. and Ont.)—History.   3. Detroit River (Mich. and
Ont.)—Environmental conditions.   I. Hartig, John H., 1952–   II. Title.

QH104.5.D48H66   2003

577.6′4′0977433—dc21   2003010283

Peter W. Stroh, 1927–2002

This book is dedicated to

PETER WETHERILL STROH
(1927–2002)

for his outstanding long-term contributions as a civic and cultural leader,
an outdoorsman and conservationist, the chairman of the Executive Committee of
the Greater Detroit American Heritage River Initiative,
a proponent of genetic engineering and molecular biology, and
former chairman of The Stroh Companies, Inc. and its Stroh Brewery.
We give thanks for his life and for his significant contributions
that will benefit future generations.

CONTENTS

Preface

Introduction

John H. Hartig

1

I

American Indians at Wawiiatanong:
An Early American History of Indigenous Peoples at Detroit

George L. Cornell

9

2

The Wyandot and the River

Kay Givens-McGowan

23

3

The Detroit River as an Artery of Trade and Commerce

John K. Kerr, W. Steven Olinek, and John H. Hartig

35

4

American Beaver Exploitation for European Chic

John H. Hartig

49

5

Waterborne Disease Epidemics during the 1800s
and Early 1900s

John H. Hartig

59

6

The Public Outcry over Oil Pollution of the Detroit River

John H. Hartig and Terry Stafford

69

7

Detroit's Role in Reversing Cultural Eutrophication
of Lake Erie

Jennifer Panek, David M. Dolan, and John H. Hartig

79

8

Mercury and PCB Contamination of the Detroit River

Jennifer Read, Doug Haffner, and Pat Murray

91

9

Lessons from Sentinel Invertebrates:
Mayflies and Other Species

Jan J. H. Ciborowski

107

10

Setting Priorities for Conserving and
Rehabilitating Detroit River Habitats

Bruce A. Manny

121

Contents

II

Biodiversity of the Detroit River and Environs:
Past, Present, and Future Prospects

James N. Bull and Julie Craves

141

I2

Preventing Toxic Substance Problems through Design and
Industrial Control of Contaminants at Their Source

D. C. Steinmetz and D. P. Thiel

171

I3

Watershed Planning and Management:
The Rouge River Experience

Catherine J. Bean, Noel Mullett, Jr., and John H. Hartig

185

I4

The Greater Detroit American
Heritage River Initiative and the Future

John H. Hartig

199

Contributors   225
Index   231

PREFACE

Webster's dictionary defines honor as high regard or great respect given, received, or enjoyed. It is time for us to honor the Detroit River. Our Detroit River is a priceless transportation corridor that has blessed us with untold economic prosperity and endless hours of contemplation and entertainment. Indeed, she is the cornerstone not just of our economy, but our very way of life.

As a boy, I often walked along the Detroit River and witnessed first-hand the destruction and erosion of her might and natural resources. In the early 1960s, the river was awash in oil and raw sewage, duck kills littered the banks, and its beaches were closed. It was the height of water pollution. While I have been in Congress memories of her past glory never left my mind, so I helped convene a landmark conference in 1962 on the rampant pollution of the Detroit River and Lake Erie. That conference changed the direction of water pollution control efforts and resulted in vast improvements and lessons documented in this book. But as the book highlights, we still have a long way to go.

The Detroit River, like many rivers located in industrial centers around the Great Lakes, has lost more than 95 percent of its coastal wetland habitats, and despite the importance of such lands, these habitats continue to be destroyed and degraded. There is now a great urgency to protect our few remaining high-quality wetlands before they are lost to further development, as well as to rehabilitate and enhance degraded ones. This will be essential to sustain the quality of life enjoyed by so many living along the Detroit River corridor.

Throughout my congressional career I have tried to carry on the legacy of conservation begun by my father, John D. Dingell, Sr., who helped create one of our nation's premier conservation laws, the Federal Aid in Sportfish

Congressman John D. Dingell

Restoration Act, commonly known as the Dingell-Johnson Act. The act provides revenues to states for conservation of lakes, rivers, and streams based on excise taxes paid for fishing tackle and equipment.

Since 1969, I have been a member of the Migratory Bird Conservation Commission, which plays a key role in the creation and maintenance of our nation's wetland and upland habitats for ducks and other migratory birds. The commission's actions have resulted in millions of acres of new conservation habitat within the National Wildlife Refuge System administered by the U.S. Fish and Wildlife Service.

I also recently realized one of my long-held dreams, the enactment of the Detroit River International Wildlife Refuge Establishment Act, the first International Wildlife Refuge in North America. I authored this act to help bring people closer to the river and to conserve its many boundless resources. In the past one hundred years, America's National Wildlife Refuge System has protected hundreds of wild species, including our national symbol, the bald eagle. The refuge system also provides Americans with tremendous outdoor recreational opportunities—including fishing, hunting, environmental education, wildlife observation, bird watching, and photography—making them convenient and accessible places for Americans to connect with nature.

It is my hope that the work of the American and Canadian Heritage River Initiatives, and the growth of the International Wildlife Refuge, will improve the quality of life for all those who are touched by this great river. Further, it is my hope that the uses and benefits of the river are sustained and enhanced for future generations. American ornithologist John James Audubon dedicated his life precisely to this goal. Audubon said, "A true conservationist is a man who knows that the world is not given by his father, but borrowed from his children." It is in that spirit that we must work to protect our Detroit River for our children and grandchildren.

Honoring Our Detroit River, Caring for Our Home tells, for the first time, the environmental history of the Detroit River and clearly articulates how we must see ourselves as part of this ecosystem and treat it like our home. The collection of stories and lessons taught in this book should be essential reading for all people in southeast Michigan and southwest Ontario. This book presents numerous examples of how our lives, our resources, and our futures are intimately and ultimately inseparable. I commend this book to you and hope it will elicit your personal commitment to honor our Detroit River and help care for our home.

INTRODUCTION

John H. Hartig

From a scientific perspective, rivers are an integral part of land. All land is part of a drainage basin or watershed. Water flows over the land and through it. McCully (1996) describes the interrelationship between rivers and the land within their watersheds as follows:

> Indeed, rivers are such an integral part of the land that in many places it would be as appropriate to talk of riverscapes as it would be of landscapes. A river is much more than water flowing to the sea. Its ever-shifting bed and banks and the groundwater below, are all integral parts of the river. Even the meadows, forests, marshes and backwaters of its flood plain can be seen as part of a river—and the river as part of them. A river carries downhill not just water, but just as importantly sediments, dissolved minerals, and the nutrient-rich detritus of plants and animals, both dead and alive. (8)

Watersheds are ecosystems composed of a mosaic of land uses connected by a network of streams (Pacific Rivers Council 1993). A watershed starts at the highest point of elevation within the drainage basin. Snowmelt and rainfall wash over and through the high ground into rivulets that drain into headwater streams. As headwater streams descend, tributaries and groundwater contribute flow and they become rivers. In the lower portions of the watershed, rivers slow and start to meander and braid, seeking the path of least resistance across widening valleys, whose alluvial floor was laid down by millennia of sediment-laden floods (McCully 1996). Eventually, all rivers flow into a lake or an ocean.

From an anthropocentric perspective, rivers have played a key role in the history of civilization. Rivers provided some of the earliest transportation routes. It is often said that rivers served as the first natural highways. Aboriginal people settled along rivers to have easy access to drinking water and hunting and fishing grounds. Indeed, there is much archaeological evidence that hunter/gatherer societies settled along rivers and in river valleys to reap their benefits. As civilizations developed, rivers continued to play critical roles in providing:

- municipal and industrial water supplies
- energy production (both hydroelectric power and cooling for thermoelectric generating stations)
- irrigation
- flood control
- transportation
- commercial fisheries
- recreation (swimming, fishing, hiking, etc.)
- assimilation of waste

These same rivers not only provided beneficial uses to cities, towns, and industries, but they were also frequently misused as they carried away human and industrial wastes as their wetlands and flood plains were filled and as their watercourses were permanently altered. In a comprehensive review of the degradation and management of river ecosystems, the Pacific Rivers Council (1993) concluded that almost every segment of society has been adversely affected by the degradation of America's rivers and we all pay heavy direct or indirect ecological, financial, and job-related costs, whether we realize it or not. Therefore, there is an urgent need to rehabilitate and protect rivers to meet the needs of our present generation without compromising the ability of future generations to meet their needs.

Caring for Our Rivers as Our Home

What we need is a new attitude about our rivers and their watersheds. A river is part of a larger ecosystem that we are also part of. We must remember that what we do to the river ecosystem we do to ourselves. There is a profound difference between environment and ecosystem. Environment is the physical and chemical setting in which life exists and is often thought of as something separate from animals and plants. Life, including humankind, is considered part of an ecosystem. It is similar to the difference between house and home. A house is something external and detached; a home is something we see ourselves in even when not there. Again, we must see ourselves as part of our river ecosystems and we must take care of them as our home. Rivers are not something to be taken for granted. We must come to see ourselves in a symbiotic relationship with our rivers. This relationship is reflected in a quote by

Mark Twain: "The face of the water, in time, became a wonderful book—a book that was a dead language to the uneducated passenger, but which told its mind to me without reserve, delivering its most cherished secrets as clearly as if it uttered them with a voice. And it was not a book to be read once and thrown aside, for it had a new story to tell every day."

The International Rivers Network (1999) is a champion of river stewardship and has provided the following eloquent description of the role of rivers in our lives:

> A river is a thing of grace and beauty, a mystery and a metaphor, a living organism whose processes have been perfecting themselves through the ages, shaping our landscapes into works of art greater than those found in any museum. Rivers feed us physically and spiritually. They determine where we live, what we eat, what we drink, and where we dance. We write songs, stories, and poems about them. We go to them in order to learn about ourselves. They provide a place of meditation, a place for celebration. (1)

There are many great rivers throughout the world, including the Euphrates, the Nile, the Ganges, the Mississippi, and the Amazon. This book is about another great river called the Detroit River. The Detroit River was created some 14,000 years ago during the retreat of the Wisconsin Glacier (figure 0.1). The present river channel was established when falling water levels permitted erosion of the lake plain and moraines.

The Detroit River is a shared resource between Canada and the United States in the heart of the Great Lakes Basin Ecosystem and along the world's longest undefended border. It is a connecting channel linking the upper and lower Laurentian Great Lakes (figure 0.2). Indeed, the entire upper Great Lakes flow through the Detroit River, making it truly unique in water quantity and other features (table 0.1). It should be no surprise why people and major chemical, steel, automotive, and ship building industries located along the Detroit River—plentiful water supplies, outstanding transportation routes, and access to minerals and resources.

The story of the human uses and abuses of the Detroit River is a powerful one, one that everyone can learn from. This book is not a comprehensive history of the Detroit River but an overview of different eras of the environmental history of the river and a call for individuals to care for their river as their home. The book begins with the Native American history along the Detroit River prior to European exploration and settlement. The environmental history of the Detroit River continues with perspectives on the river as an artery of trade and commerce, beaver exploitation, waterborne disease epidemics, oil pollution, phosphorus enrichment and eutrophication,

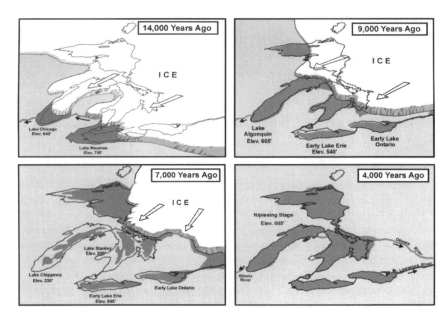

Figure 0.1. How the Great Lakes and the Detroit River were formed (U.S. Army Corps of Engineers and Great Lakes Commission 1999; graphic courtesy of U.S. Army Corps of Engineers, Detroit).

toxic substances contamination, and concern for loss of habitat and biodiversity. The book concludes with chapters on pollution prevention, watershed planning and management, and the current efforts to achieve balanced and continuous community, economic, and environmental progress under the Greater Detroit American Heritage River Initiative and the Canadian Heritage River Initiative.

The Detroit River was one of fourteen river systems in the United States designated as an American Heritage River in 1998. The Greater Detroit American Heritage River Initiative is a community-based program designed to promote environmental stewardship, encourage economic development, and celebrate history and culture. This book was written to help celebrate:

- the three-hundredth anniversary of the European founding of a settlement in the Detroit-Windsor area in 1701
- the 2001 designation of the Detroit River as a Canadian Heritage River, making it the first international heritage river in the world
- the 2001 designation of the lower Detroit River as the first International Wildlife Refuge in North America

Figure 0.2. An aerial photograph of the Detroit River, a key connecting channel linking the upper and lower Great Lakes (U.S. Army Corps of Engineers, Detroit).

TABLE 0.1
Examples of Key Features and Characteristics That Make the Detroit River Unique

- The Detroit River is 32 miles long and 0.4–2.0 miles wide.
- Water depth is variable, ranging from 3 to 50 feet; the shipping channel is routinely dredged to a depth of 27 feet for navigational purposes. The mean discharge rate is 185,000 cubic feet per second (5,240 cubic meters per second); current speed is 1–3 feet per second.
- The Detroit River contains 21 islands.
- The Canadian portion of the watershed is 90 percent agricultural, while the remaining area consists of urban, residential, and industrial lands; the U.S. portion of the watershed is only 30 percent agricultural, while the remainder is 30 percent residential, 30 percent urban, and 10 percent industrial.
- Over 5 million people live in the 700–square mile watershed.
- During 1998 total U.S.–Canadian trade exceeded $396 billion; approximately one-third of all trade that travels by road, passes over the Ambassador Bridge.
- During 1998, approximately 22 million vehicles passed through the Windsor-Detroit Tunnel or over the Ambassador Bridge.
- Shipbuilding historically had a substantial national and international economic impact. Detroit is one of the busiest ports in the Great Lakes; 19.5 million tons of cargo were handled at the port of Detroit in 1998.
- Ship captains look forward to getting their mail from a Detroit River mail boat named the *J. W. Westcott II* that has its own zip code.
- The Detroit River and the Rouge River are both identified as Great Lakes Areas of Concern or pollution "hot spots"; ten beneficial uses of the Detroit River are impaired.
- The Michigan Department of Community Health advises that no one eat carp from the Detroit River and that certain individuals eat limited amounts of certain sizes of redhorse sucker, freshwater drum, northern pike, walleye, and yellow perch because of bioaccumulation of persistent toxic substances.
- Phosphorus control at the Detroit Wastewater Treatment Plant was the single largest factor in reversal of cultural eutrophication of Lake Erie; between the late 1960s and mid-1980s there was over a 90 percent reduction in phosphorus loading from the Detroit Plant to the Detroit River; phosphorus loading since then has been fairly stable.
- Sediments in many stretches of the river are contaminated with heavy metals like mercury, oils, and PCBs, especially along the U.S. side.
- Mayflies have returned to the Detroit River, which is a sign of improved water quality. Lake sturgeon and bald eagles are successfully reproducing for the first time in many decades.
- The Detroit River is a major waterfowl migration corridor and is recognized for its significance in the North American Waterfowl Management Plan; 29 species of waterfowl are commonly found in the Detroit River.
- The Detroit River has one of the highest biodiversities in the Great Lakes Basin and has been designated as a Biodiversity Investment Area by the United States and Canadian governments; 65 species of fish are found in the river.
- The Detroit River has an international reputation for its walleye fishery; it is estimated that walleye fishing alone brings in $1 million to Downriver's economy each spring.

- The Detroit River is the host of the International Hydroplane races.
- Marshes along the lower Detroit River have been declared part of a Regional Shorebird Reserve by the Western Hemisphere Shorebird Reserve Network.
- The International Freedom Festival fireworks display, which is one of the largest in North America, occurs each year in downtown Detroit and Windsor over the Detroit River.
- Belle Isle is a 980–acre island park located in the Detroit River within close proximity to the central business district of the city of Detroit. It was designed in 1883 by Frederick Law Olmstead, who designed New York's Central Park. It provides spectacular views of Detroit, Canada, freighter traffic, and the Ambassador Bridge.
- Historic Fort Wayne is located at a strategic location on the banks of the Detroit River. It was built in the early nineteenth century during a period of tension with the British in North America. It was used during the Civil War and both world wars. Later it became one of the largest induction centers during the Korean and Vietnam Wars.
- In 1998, the Detroit River was designated an American Heritage River to promote environmental stewardship and economic development, and to celebrate history/culture.
- In 2001, the Detroit River was designated a Canadian Heritage River, making it the first international "heritage river" system in the world.
- In 2001, the lower Detroit River was designated the first International Wildlife Refuge in North America.

The Detroit River has been a working river in the industrial heartland. It has had the reputation of being a "polluted river in the rust belt." The International Joint Commission has identified it as one of forty-two Great Lakes Areas of Concern with impaired beneficial uses. In fact, the Detroit River has the Black Lagoon, a backwater that was historically polluted with oil and grease. This contributes to its reputation as a degraded and spoiled river, a perception that is no longer accurate. Dramatic improvements have occurred in the Detroit River, and it is now an invaluable, multifaceted resource that serves as the foundation of local economies, provides numerous recreational and historical opportunities and ecological values, and enhances "quality of life." This book tells unique environmental stories of the Detroit River and highlights the significant progress in rehabilitating it.

Both the Detroit River and the Detroit-Windsor metropolitan area are unquestionably unique places. Place, like nature, is very particular. Certain plants will grow in one place but not another. To fully comprehend and appreciate the significance of a place and our relationship to it, we must learn about it, understand it, respect it, honor it, and eventually care for it as our home. Our hope is that this book helps guide people to learn more about the Detroit River and to care for it as their home.

The Detroit River is unquestionably playing a key role in revitalizing and reshaping the greater Detroit-Windsor landscape. Golding (1998) has stated it best: "To deny the river is to deny the origin of the city. To rethink the river is to discover a unique opportunity to define urban places, join neighborhoods and communities together, and reconnect us to our landscape and our history" (5).

LITERATURE CITED

Golding, A. 1998. *The Los Angeles River: Reshaping the Urban Landscape.* Los Angeles: Target Science.

International Rivers Network. 1999. *A Case for Living Rivers.* Published on-line: www .org/dayofaction/index.asp?id = /dayofaction/background5.htm/

McCully, P. 1996. *Silenced Rivers: The Ecology and Politics of Large Dams.* London: Zed Books.

The Pacific Rivers Council. 1993. Report, "Entering the watershed." Eugene, OR.

U.S. Army Corps of Engineers and Great Lakes Commission (GLC). 1999. *Living with the Lakes.* Ann Arbor: GLC.

American Indians at Wawiiatanong

An Early American History of Indigenous Peoples at Detroit

George L. Cornell

The Great Lakes Basin has been home to indigenous peoples for thousands of years. Numerous archaeological sites have been identified in the southern lakes region, some of which are located along what we call today the "Detroit River" (Fitting 1975; Halsey 1999). Obviously, the river flowing through the region and connecting two large lakes was not always referred to as the Detroit River. The French first called the place "Le d'Etroit," which literally translates to "the straits." Interestingly, one has to search far and wide to discover what the river, and place, was called by native peoples during the period of early contact between the French and groups of the Miami, Huron, Ottawa, Potawatomi, and Sauk and Fox. Unlike many other Algonquin words (i.e., "Michigan") that lace our language and regional maps, the word *Wawiiatanong* has failed to be commonly used to locate or refer to the river.

The word Wawiiatanong literally means "the place on the curve or bend" and it refers to the location of the earliest historical village of Algonquins living along the river (Roy 2001). We know that the earliest inhabitants were members of the Algonquin linguistic group and that they were probably Miami, Sauk and Fox, Mascouten, or Kickapoo. These groups were well established in the region and had scattered villages north into the country surrounding Saginaw Bay. As early as 1616, Champlain met with a group of Mascoutens to the south and west of Lake Huron. The Huron or Wyandot would not become prominent residents of the area until the middle of the seventeenth century and they spoke a dialect of the Iroquois language. The usage of the word Wawiiatanong among Algonquin speakers persisted well into the middle of the 1800s and was eventually recorded by Father Baraga in his *Dictionary of the Otchipwe Language* that was first published in 1878 (Baraga 1973).

The archaeological record and the historical record both support the notion that the Wawiiatanong region had been the site of human habitation for a long period of time. Without question, the reason for these long periods of occupation was related to the incredible resource base of the region and access to the water transportation route between what would become known as Lake St. Clair and Lake Erie. The river was the most convenient route for the transportation of agricultural goods and movements of people from one locale to another. Water transportation in dugout and bark canoes (e.g., elm bark or birch bark, which now grows much further north) was particularly important since the dog was the only beast of burden used by native peoples until the introduction of domesticated livestock by colonial entrepreneurs. The region was also the site of numerous earthwork enclosures that had been constructed prior to contact with representatives of New France. These enclosures were numerous in the region that is south of Lake Erie and then north along the Detroit River, the western edge of Lake St. Clair, and continuing north into the Thumb and Saginaw Bay. Exactly what these earthwork enclosures were used for is uncertain, but it appears that some had palisades and that people prepared food inside the constructed sites (Zurel 1999). How these sites were used is still unknown, but they may have been fortifications, ceremonial arenas, or garden areas. The early settlement patterns of the region were to change drastically with the advent of colonial enterprises in the "New World."

The French made their first appearance in the upper Great Lakes in 1618, two years before the "Pilgrims" landed at Plymouth and founded their infamous colony. Most of the native peoples in the Great Lakes region allied themselves with the French and began to trade furs through Quebec. The French began to have problems early on with the Iroquois, who resided to the south of the St. Lawrence River in what is now New York. The conflict was over access to furs and trade alliances since the Iroquois traded predominantly with the English to the east. By 1644, Algonquin tribes and the Huron were actively fighting expanding Iroquois interests in the region and prolonged warfare was inevitable (Cornell, Clifton, and McClurken 1986).

In 1648, the Iroquois Confederacy began an unforgiving war on the Huron and the Algonquin villages in the Great Lakes. The cause of the war was related to the acquisition of furs that could be traded for guns and other western European goods. The Iroquois initiated a series of wars that lasted well over fifty years, causing widespread dispersal of Huron, Chippewa, Ottawa, Potawatomi, Sauk and Fox, Kickapoo, Mascouten, and Miami villages in the Great Lakes. During the course of this internecine warfare, the Lower Peninsula was almost devoid of Algonquin interests. Villages were forced to consolidate for mutual protection and many people moved to the west coast of Lake Michigan to protect themselves from marauding bands of Iroquois warriors. The village at Green Bay attracted many refugees from the war during this time, seriously disrupting life and subsistence patterns. One of the few villages in the upper Great Lakes that could withstand the Iroquois onslaught was the predominantly Chippewa village at Sault Ste. Marie, which was bolstered by other Chippewa communities and Ottawas in the region.

Even though the Great Lakes region came under increasing pressures of the fur trade and ongoing intertribal warfare, the environment was incredibly rich in fish and mammal populations. The Wawiiatanong region supported very large populations of waterfowl, including swans, geese, and ducks, and large mammals flourished. There were numerous deer, elk, and bear in the vicinity and caribou were present at points in time. The Sauk and Fox both had elk clans and the Sauk had a buffalo clan. All of the early residents of the Wawiiatanong region had access to buffalo for hides and other uses, but it appears that the primary buffalo range was to the south and west of the region. The river and adjacent lakes provided local native populations with an incredible array of fish and shellfish resources. One can only imagine what the river system and lands were like in the period before contact with the French or British. We do have enough information, however, from archaeological sites and early observations of the region to piece together a picture of how local resources provided a solid nutritional base for indigenous peoples in the area.

Like so many other aquatic and woodland environments, the Wawiiatanong region provided access to innumerable resources that sustained a high quality of life for native peoples. During the Early Archaic Period (10,000–8,000 B.P.) and into the Late Archaic Period (5,000–3,000 B.P.) there is clear evidence that native peoples in the region were subsisting on a varied diet of large game, small game, plant products, and fish (Arnold 1977; Schott 1999; Lovis 1999; Lovis, Halsey, and Robertson 1999). The diet was composed of about one-third large game and one-third plant products during this period of time with the remainder composed of fish and small game, but there appears to have been a decrease in reliance on large game as groups became more sophisticated and settled in their regional environments. Small game and fish were a very important part of local diets, but there were fluctuations in small game populations tied to natural cycles, and of course catching fish was dictated by proximity to rivers and large bodies of water. The local environments during the Archaic periods were much wetter than they are today, and there were extensive wetlands and bogs adjacent to rivers and flood plains. This was not an aquatic environment controlled by people building dams and levees in an attempt to harness the forces of nature. These environments supported incredible numbers of migratory waterfowl, fish, and mammals. Additionally, the flood plains and rich soils were a very good medium for native agriculture, and the shallow bays and inlets along watercourses were perfect areas for wild rice to grow. All of these resources were readily available to native peoples, and they possessed the technology to harvest fish and mammals at will. The atlatl, a device created to throw short spears, was used for thousands of years in the region. Native peoples also used thrusting spears and harpoons for fishing, as well as deepwater nets and fish hooks. They also constructed weirs and fish traps to catch large numbers of fish where they could be funneled into narrow openings. Eventually, during the Middle Woodland Period (0–500 B.P.), the bow and arrow became a common hunting weapon in the region. With the bow and sharp, fluted points affixed to arrows, native peoples could kill any mammal on the land easier than they could by using the atlatl or a thrusting spear. Native peoples set an assortment of traps for mammals, the best of which was the simple but deadly snare. The snare is an incredibly efficient tool since numerous traps can be easily set and they will continue to "hunt" until a mammal walks into one of them; the hunter does not have to be present to make it work. Mammals caught in snares have the habit of fighting against the restricting cord until they are exhausted, allowing them to be easily dispatched by hunters.

It was the river, however, in the Wawiiatanong region that was at the heart of commerce and subsistence (figure 1.1). Using dugout canoes that were fashioned using fire or bark canoes, native peoples fished the rapids and backwaters and traveled to other seasonal encampments and resources. The adjacent marshes were a rich source of waterfowl and eggs, and turtles were abundant. The river, tributary creeks, and marshes were also full of muskrats, which were an important food for local populations. The river course and wetlands were also excellent mediums for wild rice where the water levels were consistent over time. Wild rice was an important foodstuff since it had the capacity to be stored for long periods of time in pottery vessels or bark containers. The difficulty, then and today, lies in keeping bugs out of the grain. In the early years this could be accomplished with bee's wax or by

Figure 1.1. An early depiction of Native American life along the river (American Museum of Natural History, Department of Library Services, Neg. No. 312686, photo by Rice).

using pine pitch or a glue fashioned from the spinal column of a sturgeon to seal the containers.

The waters around Wawiiatanong literally teemed with sturgeon during spawning runs, but the large fish were always present in deeper water. The use of harpoons for securing the large fish during the spawning runs was the favored method of hunting. It must be remembered, though, that like the use of the harpoon in whaling, the barbed spear simply served to affix the boat and hunters to the fish. The fish had to be tired out by dragging the boat along and then pulled in to be killed by lance or clubbing. Sturgeon grew to very large sizes, sometimes reaching seven or eight feet in length and weighing over three hundred pounds, and yielded immense quantities of food for the amount of energy expended. Native peoples also caught large quantities of channel catfish, also important to the native diet. Suckers were also present in large numbers and were available in large quantities during spawning runs. Other fish, such as pike, bass, perch, and walleye, could be netted with ease but were speared at specific times of the year when they were most vulnerable. The reliance of native peoples on shellfish in the Great Lakes has been generally understated. Mussels were present in large numbers and were readily procured and eaten by native populations. Of course, this was at a period of time when no one had to worry about pollution, in the modern sense, and the shellfish populations of the Wawiiatanong region were flourishing.

The aquatic environment around Wawiiatanong was also home to large populations of beaver, which were an important source of food to local populations. The beaver is not a particularly difficult animal to hunt and can be ambushed on high ground when it is involved in cutting saplings and trees for dam construction and a winter food supply. The network of beaver dams provided a system of crossings and walkways for native hunters in the lowland areas. These areas also produced numerous aquatic plants, a favored food of the white-tailed deer. Archaeological evidence has demonstrated time and time again that deer were highly valued as food and a source of hides to indigenous populations. Native peoples relied heavily on deer during the fall and winter to provide them with protein. The deer were easily snared and provided a large return for energy that was expended in their acquisition. Birds, like the passenger pigeon and grouse, were also available to native hunters. One can readily discern that fish and mammals provided an abundant source of protein and amino acids to early populations in the Wawiiatanong region (Cornell 1990).

Early evidence of gathering and agricultural practices demonstrates that native populations were eating numerous aquatic tubers and nuts, and that 1,500 years ago they were growing corn, beans, and squash, as well as sunflowers for food and oil. They were also growing tobacco, which was used socially as well as in ceremonial life (Cornell 1989). When one considers the range of plants and mammals, both large and small, and fish that were available to native peoples, it becomes obvious that a solid base of nutritional foodstuffs was a part of this rich environment. Of course, the resources in the region changed over time and the early people adapted to changing environments. Neither the climate, people, nor resources were static. Rather, indigenous peoples, like the land, continuously evolved and became more sophisticated in developing technologies to utilize local resources to ensure the well-being of the people.

There is little question that on the eve of contact with Europeans, the Wawiiatanong region was a diverse environment that had numerous predator and prey species, and abundant fish and water resources. Black bears were very common in the region as were foxes, coyotes, wolves, lynxes, and bob-cats. Even the puma or cougar inhabited hardwood forests and the transition zones between the forest and clearings. These areas, called "the edge," are the prime feeding areas of large ungulates like deer and elk. There was a balance on the land as predator and prey species interacted, lived, and died, and continued the food web that sustained native peoples.

Native peoples clearly viewed themselves as the inheritors of a bounty provided for them as a part of Creation. The oral traditions of the early Algonquins clearly state that the plants and animals, fish and waters, were a gift to "The People" from the Creator. In the order of Creation, native peoples were created last, and therefore the earlier elements of creation were responsible for sustaining them. Very simply, this belief is at the very core of native spiritual perceptions of the land and animals. The plants and animals made life possible for native peoples and they ritually appeased the land and the living beings that had to die so their families could survive. Native peoples made offerings to the plants before harvesting them for medicinal purposes and hunters prayed and fasted, purifying them before they killed game. Rituals of condolence or propitiation (apology) were routinely conducted to appease the spirits of animals that had been killed. To early native hunters, everything was food, except those things that were protected by personal taboo, yet killing had to be approached in a spiritual way to demonstrate a reverence for the sacred elements of Creation that allowed life to continue (Cornell 1994). It was a cycle of life, represented by the circle and the notion that all things are related. The circle is a universal symbol that has been used by many cultures to represent strength, unity, perfection, and continuity. Native peoples perceived the circle as the physical representation of the perfection and relatedness of Creation. When early native peoples remarked that "we are all related," they were addressing more than the close-knit family and extended clan. In reality, they were acknowledging the integrity of the animate earth and Creation and their personal and spiritual relationship to it. This perception of spiritual relatedness dictated native behaviors as they interacted with their environment.

Native peoples never thought of hunting or fishing as "sport," and pro-curing food through these activities was hard, physical work. The dog was the only beast of burden they used, and the work these canines could accom-plish was limited. Land was cleared for agricultural purposes by using fire, and garden implements for keeping unwanted plants in check were made of bone or antler tools. The people worked hard to survive, and village life cen-tered around gathering, hunting, fishing, and tending garden plots during the growing season. Large surpluses were rare, although goods were traded from community to community. Ceremonial life was woven through the activities of the year, and planting and harvests, as well as hunting and fishing success, were celebrated. The family and related kin group were the focal point for organizing human activity. Information was conveyed from generation to generation through oral traditions and the stories of the people. The stories were the people's history and they provided explanations for the existence of the people and their world. Obviously, like other people living in the distant past, native peoples struggled to make sense of their surroundings and the nature of the world. It is important to stress, however, that indigenous peoples learned from their experiences and refined techniques and technologies to make their life more secure and easier.

This is an important point, given the fact that one can read contemporary archaeological literature regarding early native populations and gain very little understanding about the people who were manufacturing tools, weapons, pottery, pipes, nets, and assorted personal charms and adornments. The people seem to get lost in the systematic classification of stone tools and pottery shards. The Wawiiatanong region, like so many other areas, was inhabited by an industrious people who loved their children and families, and who tried to make their homes secure from hunger, sickness, and possibly even violence from other groups. Interestingly, they were heavily tattooed with marks of tribal origin and distinction.

The early inhabitants of the Wawiiatanong region had an intimate understanding of the land and its resources and what they could be used for. This came about over generations of walking on the land, observation, and trial and error. The cycles and fluctuations in weather, and the change in the seasons, were closely watched. The people cataloged an incredible range of environmental knowledge that helped them find game and fish or other resources at any time of the year. They built dwellings that were more than capable of protecting them from rain and snow, and they had harnessed fire to heat themselves as well as cook their food. They had food, they had shelter, and they had water.

Native peoples have historically believed that water is the lifeblood of the earth. As everyone knows, and some dispute, native peoples in North American have unanimously viewed the earth as a living, feminine entity. She has been called Mother Earth in more recent centuries, but no matter what you call her, she is the source of sustenance for native peoples. As a living organism, analogous to the human body, the earth has blood (water) that courses through her veins (i.e., rivers, underground streams) that keeps her alive and vital. Water is life to the earth, just as blood is life to mankind. There is no doubt that native peoples clearly understood the relationship between water and life. Rains replenished the earth with water, purifying the land and helping bring forth new life in the spring. Some indigenous peoples believed that the rains that fell from the heavens, the source of masculine power personified by the sun, actually impregnated the feminine earth. The thunders announced the coming of the rains and were universally heralded as great powers by indigenous peoples. Likewise, lightning was respected as a powerful symbol of Creation. Clouds that make noise and flashes of fire from above are strong medicine. Experientially, native peoples understood the importance of water and its relationship to the health and welfare of the people and the earth (Cornell 1992).

This was especially true in the Wawiiatanong region. The big river that cut the land in two made a rich life possible for the people. The river was a powerful force that needed to be respected and propitiated. People routinely made offerings to the water and the spirits under it before traveling or fishing. The rapids in the river could be dangerous to the unskilled and at times of high water were better left untested. The natural flow and ebb of the river was constantly observed by native peoples and they behaved accordingly. The river was the source of life in the region. It provided the fish and aquatic mammals for the people. Large mammals drank from the riverbank or from creeks feeding the mainstream, and waterfowl lived in the backwaters and marshes created by periodic flooding and rains. The river was at the very center of life and existence.

The river was also a source of commerce and news from distant villages. Goods that were traded from one group to another could find their way to Wawiiatanong and be traded or given as gifts to relatives. The river facilitated the coming together of diverse clans and families for feasts and ceremonies. These social and solemn occasions also allowed young people to meet prospective mates and families to renew strong economic bonds that would prove indispensable in times of scarcity. All of these activities were undertaken along the banks of the big river at Wawiiatanong. Life was good along the banks of the big river. It was precarious at times, but generally the big river provided for the people.

It was truly a beautiful sight, sitting on the banks of the big river in the early morning. The sun rose almost directly upstream from Wawiiatanong in the middle of the summer. Waterfowl flew everywhere and the cacophony of their calls made the people smile as they glanced up to watch them pass by. The best sounds, though, were those of the adjijaaks (sandhill cranes) as they flew by in small groups or called to one another in the adjacent marshes. Their calls carried for incredibly long distances. The cranes' mating calls in the spring always brought good cheer to the people and they knew that the planting season was near. More than anything else, though, the big river provided tranquility, a sense of well-being. The river provided solace for the people and they thanked the spirits and the Creator for such a wonderful gift.

Of course, things changed with the "discovery" of the New World. Colonial undertakings for profit began to unfold in North America and native peoples became, at times, willing participants in the competition for empire in the new lands. Trade relations were established with the French and British, and early on the fur trade became a way of life to promote tribal or national sovereignty of indigenous peoples. The competition between native groups increased dramatically as the contest for furs was exacerbated. Eventually national wars, like the Iroquois conflict of the seventeenth century, became a priority for native peoples in the Wawiiatanong region. Village life was seriously disrupted with the onset of the Iroquois Wars in the winter of 1648, and the groups of Sauk and Fox, Mascoutens, Kickapoo, and Potawatomi in the region were forced to remove to safer quarters.

The Wawiiatanong region remained somewhat of a mystery to French explorers until late in the Iroquois Wars. Rufus Blanchard (1880), in *Discovery and Conquest of the Northwest*, recounts how Fathers Galinee and Dablon visited the site in 1670: "The first object of interest they beheld was a barbarous piece of stone sculpture in the human form. This was quite sufficient to unbalance the equilibrium of the two fathers, whose zeal had been whetted into an extravagant pitch by the hardships they had encountered on their way. With pious indignation they fell upon the 'impious device' with their hatchets, broke it in pieces, and hurled the fragments into the river."

One has to wonder exactly what it was that the good fathers destroyed that day. C. I. Walker (1886), in his article "Early Detroit," written in 1858, stated: "While doubtless, solitary missionaries and adventurers visited the banks of the Detroit at a much earlier date, I have been unable to find any authentic account of its being visited by white men previous to August, 1679, when the gallant La Salle sailed up our beautiful river in the ill-fated Griffin, built near Black Rock."

The general region was described by Father Hennipen, who served as the historian for the voyage. Obviously Walker was wrong about earlier visitors, but Hennipen's observations on the region that were recorded as a part of the voyage proved to be important to the historical record. Hennipen's observations (Walker 1886) included:

> On the tenth [of August, 1679] we came to an anchor at the mouth of the Streight which runs from the Lake Huron into that of Erie. The eleventh, we went further into the Streight, and passed between two small islands, which make one of the finest prospects in the world. The Straight is finer than that at Niagara, being thirty leagues long, and everywhere one league broad, except in the middle that it stretches itself, forming the Lake we have called St. Claire. The navigation is easie on both sides, the coast being low and even. It runs directly from north to south. The country between these two lakes is very well situated, and the soil is very fertile. The banks of the streight are vast meadows, and the prospect is terminated with some hills covered by good fruit, groves and forests, so well disposed, that one would think nature alone could not have made, without the help of art, so charming a prospect. That country is stocked with stags, wild goats and bears, who are good for food, and not fierce as in other countries; some think they are better than our pork. The turkey cocks and swans are there also very common; and our men brought several other beasts and birds whose names are unknown to us, but they are extraordinarily relishing. The forests are chiefly made up of walnut trees, chestnut trees, plum trees and pear trees, loaded with their own fruit and vines. There is also an abundance of timber fit for buildings, so that those who shall be so happy as to inhabit that noble country cannot but remember with gratitude those who have discovered the way, by venturing to sail upon the unknown lake for above one hundred leagues.

The early commentary on the Wawiiatanong clearly seems to extol the virtues of the region's natural resources. It was not until 1699, however, that Antoine De La Motte Cadillac began to work to establish the fort at Detroit that would become the region's first permanent French settlement. It was in that year that Cadillac visited Louis XIV's court at Versailles and proposed the creation of a permanent fort at Detroit. It was clearly an attempt by Cadillac, representing French interests in the region, to thwart Iroquois and English access to the upper Great Lakes and the Illinois Country. Cadillac reflected on the region:

> On both sides of this strait lie fine, open plains where the deer roam in graceful herds, where bears, by no means fierce and exceedingly good to eat, are to be found, as are also the savoury poules d'Indies (wild duck) and other varieties of game. The islands are covered with trees; chestnuts, walnuts, apples and plums abound; and in season, the wild vines are heavy with grapes, of which the forest rangers say they have made a wine that, considering its newness, was not at all

bad. The Hurons have a village on Le Detroit; they see, according to their needs, its advantages. Michilimackinac is an important post, but the climate will ever be against it; the place will never become a great settlement. Le Detroit is the real center of the lake country—the gateway to the West. It is from here that we can best hold the English in check. I would make it a permanent post, not subject to changes as are so many of the others. To do this it is but necessary to have a good number of French soldiers and traders, and to draw around it the tribes of friendly Indians, in order to conquer the Iroquois, who, from the beginning, have harassed us and prevented the advance of civilization. The French live too far apart. We must bring them closer together, that, when necessary, they may be able to oppose a large force of savages and thus defeat them. Moreover, the waters of the Great Lakes pass through this strait, and it is the only path whereby the English can carry on their trade with the savage nations who have to do with the French. If we establish ourselves at Le Detroit, they can no longer hope to deprive us of the benefits of the fur trade. (Burton, Stocking, and Miller 1922)

Cadillac's commentary on the region is similar to Hennepin's and probably drew heavily on the latter's impressions of Wawiiatanong.

The rest of the story is colonial and American history. Cadillac left Montreal in early June 1701 with fifty soldiers and fifty traders, and citizens of New France, and on July 24 of the same year, he landed and established the fort at Wawiiatanong. By 1705, there were over two thousand Indians of diverse nations residing in villages in close proximity to the fort. It is ironic that the presence of these "savages," as Cadillac alluded to them, secured the fort for French interests in the region.

What happened at Wawiiatanong was repeated time and time again across North America. Native peoples were displaced from the land and resources, and new cultures began to use the land and resources with a vengeance. Native peoples died in large numbers from newly introduced diseases and new cultures took possession of the country. It is an old and all too familiar story. The historic Wawiiatanong can only be summoned to mind when one sits on the banks of the big river and ponders what it must have been like. It is vitally important, though, for citizens to learn the history of the region that is now Detroit and the "big river." What became of the vast wildlife resources along the riverbanks? What became of the great sturgeon that lived in the big river? What became of the sacred land that native peoples celebrated and propitiated? How did the new cultures that took possession of the land care for it and preserve it for their children?

The modern, "civilized" legacy relating to the Detroit River is not a good one. Pollution and destruction of wildlife habitat seem to be bywords for development. Without question, nineteenth-century citizens thought that the resources of the land and water were inexhaustible, but they were wrong. Native peoples lived in the Wawiiatanong region for thousands of years and when newcomers took possession of the land the resource base had real integrity. In just a little over three centuries, the resource base has nearly been destroyed, and the big river and surrounding waters have been poisoned. Who is going to eat mussels out of the big river or Lake St. Clair today?

Of course, there have been innumerable changes during this period of time. Population growth, industrialization, and the rise of large urban areas have all affected the quality of our environments. What is important today is what we do about the future. We cannot turn back the hands of time and try and remake Wawiiatanong as it existed five hundred years ago. What we can do is care about our environment and work to restore the former integrity of the land and water. We can organize human activity to promote environmental wellness and clean water, and we can educate children to create a better world and to work toward sustaining our natural resources. We can learn to treat our river as our home.

If there is such a thing as "Spirit of Place," Wawiiatanong has it. For generations untold, native peoples raised their families along the banks of the big river and prayed for safe passage and bountiful harvests. They prayed for good hunting and for the land and the waters. They offered tobacco and burned sweet grass along the banks of the big river in the belief that their actions would please the Creator. These acts are a part of the history of the place and the Spirit of the land (Cornell 1985). People today may not remember these acts, but it does not diminish the fact that they occurred. The big river was the lifeblood of native peoples at Wawiiatanong, and it is still the lifeblood of people today. Hopefully, contemporary society and future generations can renew their sense of place, purpose, and responsibility, and reclaim a heritage that promotes good conservation and intelligent use of our most precious resource: water.

LITERATURE CITED

Arnold, J. 1977. Early Archaic Subsistence and Settlement in the River Raisin Watershed, Southeastern Michigan. Cited in Michael J. Schott, "The Early Archaic: Life after the Glaciers." In *Retrieving Michigan's Buried Past,* ed. J. R. Halsey, 1–81, Bloomfield Hills, MI: Cranbrook Institute of Science.

Baraga, R. R. 1973. *A Dictionary of the Otchipwe Language.* Minneapolis: Ross and Haines.

Blanchard, R. 1880. *The Discovery and Conquest of the Northwest.* Chicago: Cushing, Thomas and Company.

Burton, C. M., W. Stocking, and G. K. Miller. 1922. *The City of Detroit, Michigan, 1701–1922.* Detroit: S. J. Clarke.

Cornell, G. L. 1985. "The Influence of Native Americans on Modern Conservation." *Environmental Review* (special issue) 9.2 (summer): 104–17.

———. 1989. "Unconquered Nations: The Native Peoples of Michigan." In *Michigan: Visions of Our Past.* East Lansing: Michigan State University Press.

———. 1990. "Native American Perceptions of the Environment." *Northeast Indian Quarterly* 7.2 (summer): 3–13.

———. 1992. "American Indian Philosophy and Perceptions of the Environment." Report 140. In *Human Values and the Environment.* Madison: Wisconsin Academy of Sciences, Art and Letters.

———. 1994. "Native American Perceptions of the Environment." In *Buried Roots and Indestructible Seeds: The Survival of American Indian Life in Story, History and Spirit,* 21–41. Madison: University of Wisconsin Press.

Cornell, G. L., J. A. Clifton, and J. M. McClurken. 1986. *People of the Three Fires.* Grand Rapids, MI: Michigan Indian Press.

Fitting, J. E. 1975. *The Archaeology of Michigan.* Bloomfield Hills, MI: Cranbrook Institute of Science.

Halsey, J. R. 1999. *Retrieving Michigan's Buried Past.* Bloomfield Hills, MI: Cranbrook Institute of Science.

Lovis, W. A. 1999. "The Middle Archaic: Learning to Live in the Woodlands." In *Retrieving Michigan's Buried Past,* ed. J. R. Halsey, 83–94. Bloomfield Hills, MI: Cranbrook Institute of Science.

Lovis, W. A., J. R. Halsey, and J. A. Robertson. 1999. "The Late Archaic: Hunter-Gatherers in an Uncertain Environment." In *Retrieving Michigan's Buried Past,* ed. J. R. Halsey, 95–124. Bloomfield Hills, MI: Cranbrook Institute of Science.

Roy, H. Telephone interview, October 16, 2000.

Schott, M. J. 1999. "The Early Archaic: Life after the Glaciers." In *Retrieving Michigan's Buried Past,* ed. J. R. Halsey, 71–82. Bloomfield Hills, MI: Cranbrook Institute of Science.

Walker, C. I. 1886. "Early Detroit." *In Report of the Pioneer and Historical Society of the State of Michigan.* Lansing, MI: Thorp and Godfrey.

Zurel, R. L. 1999. "Earthwork Enclosure Sites in Michigan." In *Retrieving Michigan's Buried Past,* ed. J. R. Halsey, 244–48. Bloomfield Hills, MI: Cranbrook Institute of Science.

THE WYANDOT AND THE RIVER

Kay Givens-McGowan

Three hundred years ago, Wyandots and other native people had already inhabited the Detroit area for one millennium. The flat pancake of Wayne County was once almost all bog and marsh. The Huron River was a great drain. Gibraltar was a delta. Beaver and their dams occupied over a thousand square miles. The original people of Detroit spent their lives on the edge of waters teeming with wildlife of every kind. Living on the Detroit River was an earthly paradise.

Contrary to recent shortsighted celebrations, three hundred years ago when the Wyandots brought Cadillac by canoe the length of Lake Huron from Fort Mackinaw, it was not to discover Detroit. It was to see if what the natives said about the abundance of the region could possibly be true. Cadillac wanted to know if the beaver were as plentiful as the Wyandots claimed. Beaver was the rage in France, and the beaver trade would make French fur traders rich.

The Wyandot Indians had lived on the Detroit River for centuries. The river was their home. They fished, swam, raised families, built canoes, gathered cranberries in the bogs, and played their games on this river they loved.

The story of the Detroit River and the story of the Wyandot are one and the same, but much of the Wyandot story has been "written out" of Michigan history. The Wyandot were not here to tell their story since most of their tribe was forcibly removed from Michigan. Others kept their heads down as there was no advantage to being identified as an "injun."

The French called the Wyandot (Wyandotte, Wiandotte, Wendat, and Quendake) the "Huron" because the first Wyandot they encountered had a unique hairstyle. Their heads were shaved except for a strip down the middle, which stood straight up (with the oil of bear grease). When the French saw them, they were amused by the hairstyle and responded by calling them "Hures," which meant "wild boars." The word "hure" was also seventeenth-century French slang for "uncouth" or "backward." When anglicized it became "Huron." The present-day Wyandot of Anderdon prefer to be called "Wyandot."

The Wyandot had the misfortune of being on the losing side in the wars Europeans fought in this region and found themselves powerless after being on the losing side three times in a period of 125 years.

In 1842, the Wyandot from Michigan, Ohio, and Canada were forcibly removed to Indian Territory in what is now Oklahoma and Kansas. The Ohio Wyandot, who had settled on the south side of Lake Erie in present-day Sandusky where Chief Orontony's band established villages around 1736–38, were removed from Ohio in July 1843 because they were granted a stay of one year. Most of them were taken west.

Place names, the early archaeological record, and oral history show that the Wyandots had strong ties to the Detroit River. The story of their relationship with the Detroit River is a powerful one.

Wyandot Villages on the Riverfront

The Wyandot Nation lived along the Detroit River on both sides—present-day Canada and the city of Detroit (figure 2.1). The Wyandot name for Detroit was "Oppenago"—"the place where the waters meet." The Wyandot were slash-and-burn agriculturalists who relied on the "three sisters" for survival—corn, beans, and squash. Every ten to twelve years when the soil became depleted, they moved their villages to a location nearby (Demeter 1993, 3).

The present-day cities of Ecorse and River Rouge had Wyandot and Potowatomi villages side by side on the riverfront. The Wyandot village was called "Tonquish." The Potowatomi village was called "Robiche."

Just a few miles south of Tonquish, the Wyandot lived at the village of Maguagon, which today is the city of Wyandotte. Present-day Bob-Lo Island, called Bois Blanc by the French, was named "Atieeronnon" by the Wyandot. In Iroquois, the language of the Wyandot, Atieeronnon means "the place of the white ash tree people." Wyandot villages stretched from the St. Clair River down the Detroit River all the way to Monroe. Monroe was called "Numma Sepee" or "the place of the Sturgeon" (Zeisler 1969, 8).

Figure 2.1. The Splitlog Band (1880) of the Wyandot of Anderdon Nation (Lil Splitlog).

Wyandot Burials

The Wyandot Indians had a unique form of burial that makes their burial grounds easy to distinguish from other tribes. The large burial pits of the Wyandot are called ossuary burials. There were two burials for each person, the first occurring three days after death. The body was wrapped tightly in the deceased's finest fur skins or robes (beaver was the preferred burial skin). The body was placed on the reed mat he or she had died on. The reeds came from the river's edge. The face of the deceased was painted red and black. The family would "keen" for the dead. Community members would prepare a feast. Relatives and friends from faraway villages would arrive bearing gifts to honor the deceased. The funeral would begin at sunrise on the third day. Four men carried the body to the individual burial site. The deceased was placed in a bark coffin on a platform four feet from the ground. The mourning period for close family members lasted one year.

The second burial took place ten to twelve years later. All of the people who had died during that period would be buried a second time in a large ossuary pit called "Yandatssa" ("cooking kettle"). The spirits of the dead were symbolically "cooked" together, which released the spirits of all those who had died in the previous decade. The "kettle" ceremony was accompanied by the "The Feast of the Dead." This was a happy time in Wyandot society because it meant the long period of mourning was over.

The second ceremony confirmed the release of a Wyandot's second soul. The first soul had left the remains at the time of a person's death and traveled to the western door or the spirit world on the wings of a sacred red-tailed hawk. The feast of the dead meant the second soul had also reached the spirit world. This was cause for celebration.

There is archaeological evidence of the Wyandot's ossuary burial pits up and down the river. Along the Detroit and St. Clair Rivers there are designated archaeological sites where Wyandot ossuary pits are found—among them are the Bussinger, Riviere au Vase, Gibraltar, Younge, Missionary Island, and the Libben sites (Stothers 1999, 194–210).

The Wyandot always located their ossuary pits across a body of water from their villages. They believed the first soul could not travel across water to snatch the person's second soul before the feast of the dead occurred. The two burials were required for a person to rest forever in peace.

The archaeological Younge site, at the western basin of Lake Erie, had at least three longhouses, the common house type of the Wyandot (and all the Iroquois of the northeast). Skulls with holes drilled into them were found, which also suggested a Wyandot burial. Several days before final interment, the mummy bundles of the Wyandot (from their first burial) would have been unwrapped and the remains would have been put on display in the family's longhouse. This could have required holes in the skulls to mount the remains on the display poles. Since the Wyandot were the only people who practiced this burial type, one could conclude these were Wyandot ossuary burials on the Younge site.

The life spirit that lives in every person was called the Aki soul. The second soul, which was released at the feast of the dead, was called the Asken. The bones of the dead were called Aatisken and were sacred.

Wyandot people who knew they were dying often painted their face so they would look beautiful when they went to greet their relatives in the Western Door (the Spirit World), the Village of the dead. This paint was found at the mythological Big Rock. Present-day Brownstown and the city of Gibraltar were named "Toh Roon Toh" (Big Rock) by the Wyandot. There was a huge boulder at the edge of the Detroit River at Gibraltar (Babcock 1837; Knopf 1960; Michigan Pioneer and Historical Society 1895; Lowrie and Clarke 1832). This Big Rock called "Ekarenniondi" was spiritually and symbolically associated with the Big Rock in Gibraltar. Big Rock is also the site of an ossuary burial of the Late Woodland (A.D. 600–1000) people and the Wyandot of long ago, indicating this area had always been a spiritual place to native people.

Village Site in Present-day Gibraltar

The early site of Toh Roon Toh was chosen as the site for a Wyandot village based on the spirituality of the Wyandot and because of a creek and a turtle pond there, which made this site special. The Wyandot are the children of the turtle. Five of their twelve clans were named for various kinds of turtles.

The village of Chief Adam Brown, "Tahounehawietie" in Iroquois, was the site of several historical Indian councils during the 1700–1800s. The Wyandot were the "Keepers of the Council Fire," a prestigious and powerful position in native society. The "United Indian Nations" council was held at Brownstown in 1786 (Michigan Pioneer and Historical Society 1888).

The upper Great Lakes of Lake Huron, Lake Michigan, and Lake Superior are connected to the lower lakes of Lake Erie and Lake Ontario by three important bodies of water. Lake St. Clair, the St. Clair River, and the Detroit River were important transportation routes and village sites of Late Woodland native people and all Michigan's native people.

Wyandot Spirituality Connected to Economic Lifeways

Everything in nature was linked as binary opposites, which also linked the twelve matriarchal clans of the Wyandot to everything that lives. Fish were linked to the river, fields were linked to mountains, air was linked to the sky, as the owl was linked to the eagle. This interconnectedness of all things was considered a sacred bond.

The Wyandot connection to the Detroit River was also a sacred bond. They believed that the Spirit that lived in the fish was called "Oki." In each fish house there was a medicine man skilled in talking to the fish, who would tell the fish that they were respected. Respect would be shown to the fish by the burial of their bones rather than burning them. The Wyandots considered the burning of bones, human or animal, to be a disrespectful act. Medicine men would thank the fish in advance for giving up their lives so that people could eat and live. Native people believe that all things must die so that something else can live. The way a living being dies and the way in which its body is treated after death are important.

Water also figured in the spiritual lives of the Wyandot. Tobacco would be placed on the water as an offering to the spirits that lived in the water, which was a way of thanking the water for giving people life. The water, it was believed, gave life to all things. Water represented life and purity. There were many ceremonies involving the use of water and honoring of this precious resource.

Fishing was done in a variety of ways, either with a spear or with fishhooks and line, which were used for small fish. A barbed head that had been whittled and ground out of bone was attached to a wooden spear for spearfishing. Hooks were made from wood with a bone barb attached. Sturgeon bones were often used in making the barbed heads on spears and fishhooks.

Fishing was also done with nets on the open water. Fish fences or weirs were often placed where creeks or streams emptied into the larger river. The Wyandot villages were near the creeks or streams because the fishing weirs guaranteed a constant food supply. Otter Creek and Stony Creek in Monroe were both village sites.

In October and November, the Wyandot fished for Atsihiendo, or lake trout. In the spring they fished for walleye, whitefish, bass, and sturgeon. All fish were called "Leinchataon."

Fishing was also done in the Detroit River with nets. Each fishing party had its own leader. The fishing parties would build small bark cabins on the islands in the river when the fish would be spawning.

The fishing parties set their nets in the evening and pulled them up in the morning. Once the fish were pulled up they were immediately gutted (except for freshwater cod called burbot) and spread on racks to dry in the sun. Some of the fish would be smoked over a fire and then packed into bark containers to be eaten all year-round.

Fish were also boiled for the oil, which was used as a seasoning for sagamite and other corn dishes. The Wyandot also mixed the oil with paints for painting their bodies.

Burbot was used as a flavoring in corn soup, a Wyandot favorite. They were caught from nets cast from the shore. They were hung up to dry in bunches beneath the longhouse roofs. Burbot could then be eaten all winter. Children were given cod liver oil to drink because the Wyandot believed it was good for their health.

The Wyandot traveled back and forth across the Detroit River in their large canoes made only by the Sea Snake Clan. Smaller canoes for personal use were often obtained through trade with the Potowatomi. These two peoples, Wyandot and Potowatomi, had a reciprocal relationship based on need and friendship. Travel across the Detroit River from the Canadian side to the island of Grosse Ile was common.

Catholicism and the River

Many of the Wyandot had converted to Catholicism and went to the mission chapel on Bois Blanc until it was burned down, and later to Detroit. Both the mission of St. John the Baptist in Amherstburg (1802) and the old mission of L'Assumption in the Township of Sandwich, Ontario (1790s) were also Wyandot parishes. In the late nineteenth century many Wyandot Catholics would pull up their canoes at the foot of present-day Church Street on Grosse Ile to hear mass at St. Anne's chapel on Sundays and Holy Days. The Wyandot Green Corn Ceremony, held the first week in August, brought many Catholics from across the river back to St. John the Baptist. Church records indicate that it was a common practice for Wyandots to bring their babies back across the river to be baptized.

The Wyandot called the north stretch of the Detroit River Tsugsagron-die, meaning "where beaver is plentiful." They called the lower half of the river that included Teuchsay-Grondie (Grosse Ile) and the ten islands chain Erige Tejocharontiong, which means "the place of the Erie people where there are many beaver dams" (Zurel 1961, 10).

Wyandot Games

The frozen Detroit River became a playground for the Wyandot. Father Gabriel Sagard, who arrived with the French fur trappers, was the first person to document a game still popular today—hockey. Father Sagard was a Jesuit missionary who worked among the Wyandot Indians from 1625 to 1650. He wrote annual reports to the superiors of the Jesuit order in France, which were published as *The Jesuit Relations and Allies Documents.* These books give the earliest written accounts of the lives of native people in the Americas.

Sagard and other Europeans often reported on Native American game playing, calling it "excessive" or "brutal." They failed to recognize that many skills that native people valued—endurance, speed, strength, stamina, agility, and intuition—were taught through these games. The games of hockey, lacrosse, snow snake, skinny, and double ball were all played by the Wyandot. The Wyandots' version of hockey was very rough and no padding was worn. Father Sagard reported many broken legs and arms. He reported that boys and young men "play a game with curved sticks, making them slide over the ice and snow, and they hit a ball of light wood." Sagard was fascinated with the game, as he had never seen it played on the river. Even the game of hockey tied the Wyandot to the river in a symbolic relationship.

Warfare among the Tribes

The Wyandot were closely related to the Erie Nation, who spoke Iroquois. There was a great deal of intermarriage and interaction between the two people. The Erie, like the Wyandot, were decimated by the warfare with the Iroquois that lasted from 1649 to 1700. The warfare was a result of the Dutch West Indies Company's arming the Iroquois with guns and sending them out against the Tionnontati, Neutrals, Nipissings, Chippewa, Ottawa, Potowatomi, Wyandot, Erie, and Susquehanna. The Dutch wished to control the fur trade after depleting the numbers of beaver, fox, and other furbearing animals in the eastern half of the northern United States.

The remnant band of the Erie asked to be adopted by the Wyandot because they knew it would be difficult to survive with fewer than two hundred people. The Wyandot adopted the entire remaining Erie in 1653. The Erie dropped their names and took Wyandot names at the adoption ceremony. Today the name of Lake Erie is our reminder of the tribe once known as the Erie. The Wyandot territory then came to include the region on the south side of Lake Erie in present-day Ohio, which had been Erie territory.

During the French and Indian Wars of 1689–1763, the Wyandot sided with the French, against the British. The British eventually won and Wyandot territory then became British territory.

During both the American Revolution and the War of 1812, the Wyandot sided with the British The Wyandot found themselves on the losing side of three wars.

Wyandot land claims, both at Brownstown and at the U.S. Treaty of Fort Harmar, had recognized Monguagon (Wyandotte) and Brownstown (Gibraltar and Brownstown Township) villages in 1789. These claims to the villages were later inadvertently rescinded when the Wyandot chiefs from Detroit arrived late for the signing of the Treaty of Greenville in 1795 (Burton 1908, 101). "Mad" Anthony Wayne ("Blacksnake" to the Wyandot) later agreed to give back to the Wyandot the important Detroit River villages. But Wayne never put this agreement in writing and he died the next year (Fuller 1928a; Fuller 1928b; Jefferson 1809; Greeley 1810; Walker 1816). No one knew of his agreement except for the Wyandots.

The Wyandot were angry over the loss of their two largest villages on the Detroit River and petitioned Congress and the president on February 28, 1812, asking for reconsideration to terms that had been put forth by U.S. government. Chief Walk-In-The-The-Water of Monguagon and Chief Roundhead of Brownstown went to Washington, D.C., to speak with President James Madison about getting their villages back (Hathon 1836; Bland 1989, 6). Secretary of War Henry Dearborn was against the idea and the Wyandot lost all hope of having their villages of the Detroit River returned to them. Angry and worried about basic survival of the Wyandot Nation, the chiefs returned to Michigan. The consensus was to fight alongside the British after the betrayal by the Americans.

Brownstown's position, directly across from Amherstburg, was strategically located for the British and their Wyandot allies (Carter 1934, 125). The British could bring their troops and supplies across the frozen ice to Brownstown in winter. The Wyandots could use their large seagoing canoes to move men across to Amherstburg or blockade the Detroit River when necessary. U.S. General Hull had built a supply road from Ohio north to Detroit. The Wyandots and their Indian allies successfully blocked the supply line to Detroit for two weeks.

The great war chiefs Roundhead and Splitlog led the Wyandots, and Tecumseh, the great chief of the Shawnee, and his warriors had arrived at Brownstown to aid the Wyandot. The people of Frenchtown (present-day Monroe) had been under pressure to join with the Wyandot against the Americans. Chiefs Roundhead, Splitlog, and Walk-In-The-Water sent this message to the people of Frenchtown: "Friends, Listen! You have always told us you would give the Wyandot any assistance in your power. We, the Chiefs of the Wyandot, call upon you all to rise up, and bring your arms along with you. Should you fail this time, we will not consider you in the future as friends, and the consequences may be very unpleasant" (Bland 1989, 6). Despite these pleas, the Frenchtown citizens sided with the Americans.

In July 1812, over five hundred chiefs and warriors of the Midwest Indian nations held their last great council at Brownstown on the Detroit River. The Wyandot remained in Brownstown only one more year until 1813. Public sentiment was building against the Wyandot. They not only had been merciless in warfare, but also white men wanted their valuable land along the water on both sides of the Detroit River. The Wyandots were removed from their beloved Detroit River in 1816 and were placed, along with some Shawnee, Potowatomi, and Ottawas living with them, on a reservation along the Huron River in Flatrock. Even though they were living on a river, they cut a path back to the Detroit River that is present-day Sibley Road (Garriston 1897, 16–25).

Removal to the West

By 1818, all of the prime waterfront land of the Wyandot had been lost to white settlers. During the removal period, thirty-four tribes were forcibly taken from the east to Indian Territory in present-day Oklahoma and Kansas. This tragic period is most poignantly recalled by "the trail of tears," which was a genocidal removal of the five large tribes of the southeast. The Wyandot and other tribes walked their own trail of tears.

The removal in 1842 took most of the Wyandot to Kansas (and twenty years later to Oklahoma). The U.S. government had promised the Wyandot 160,000 acres on which to resettle. But the Wyandot who survived the forced march to Kansas found no land waiting. The Delaware people gave the Wyandot thirty-six sections of land because they recalled the Wyandot letting them "lay their blankets down" more than a century earlier when the Delaware had been driven from the east and found refuge in Wyandot territory in Ohio.

Wyandot in Michigan Today

There was a small group of Wyandot who were not driven to Kansas. They hid on the islands, fled to Canada, and later returned to Michigan. Some Wyandots were allowed to stay if they had "legal" title to land. This group remained in Michigan and is the Wyandot of Anderdon Nation. Today the Wyandot of Anderdon, who have remained on the river, continue to view the river as their ancestors did—a sacred place to be honored, respected, and used by everything that needs clean water to live. Today many still live along the Detroit River

There are over nine hundred members of the Anderdon Tribe of Wyandot living in southeastern Michigan, many in the Downriver Detroit community that they occupied for a millennium. They belong to almost every hunting and fishing club in the region. The membership rosters of environmental organizations, including the Friends of the Detroit River, are full of Wyandots.

The Wyandot use of the water included not only transportation, but a *lifeway*. The Wyandot canoed across the river to have children baptized, fished for their favorite fish from the river, buried their ancestors next to the river, and gathered on the water's edge for celebrations and ceremonies.

The Wyandot connection to the water was a spiritual one. They recognized that their lives were dependent on the Detroit River. As a result, the Wyandot had the deepest respect for the water and probably the broadest use of this valuable resource of any people who have ever lived along the Detroit River.

LITERATURE CITED

Babcock, J. 1837. Plat of Gibraltar. Wayne County, Michigan. Photocopy maintained at Tax Department, Gibraltar, Michigan.

Bland, J. 1989. *Our Great Chiefs*. Wyandotte, OK: Historical Committee, Wyandotte Tribe of Oklahoma. 6–12.

Burton, C. M. 1908. "Prefatory Explanation." *Michigan Pioneer and Historical Collections* 36:101–11.

Carter, C. E. 1934. *The Territorial Papers of the United States: The Territory Northwest of the River Ohio, 1787–1803*. Washington, DC: U.S. GPO. 125–56.

Demeter, C. S. 1993. Phase 1 Archaeological Survey in Gibraltar. Site 20WN9. Wayne County, Michigan. 3–30.

Fuller, G. N., ed. 1928a. *Geological Reports of Douglass Houghton, First State Geologist of Michigan 1837–1845*. Lansing, MI: Michigan Historical Commission.

———. 1928b. *Documents Relating to Detroit and Vicinity, 1805–1813*. Michigan Historical Collections, Vol. 40. Lansing, MI: Michigan Historical Commission.

Garriston, F. 1897. *Flatrock of 60 Years Ago*. Flatrock, MI: Flatrock Historical Society. 16–25.

Greeley, A. 1810. *Plan of Private Claims in Michigan Territory*. Detroit: Burton Historical Collection, Detroit Public Library.

Halsey, J. A. 1999. *Retrieving Michigan's Buried Past: The Archaeology of the Great Lakes State*. Bulletin 64. Bloomfield Hills, MI: Cranbrook Institute of Science.

Hathon, A. E. 1836. Section 354 and 355 at Brownstown owned by Lewis Cass. Deed Liber 14:301. Wayne County Registrar of Deeds, Detroit.

Jefferson, T. 1809. *A Declaration of Behalf of the Wiandottes Near Detroit*. Thomas Jefferson Papers, Burton Historical Collection, Detroit Public Library.

Knopf, R. C., ed. 1960. *Anthony Wayne, A Name in Arms: The Wayne-Knox-Pickering-McHenry Correspondence*. Pittsburgh: University of Pittsburgh Press.

Lowrie, W., and M. St. Clair Clarke. 1832. *American State Papers: Indian Affairs*. Class 2, vol. 1. Washington, DC: Gales and Seaton.

Michigan Pioneer and Historical Society. 1888. Haldimand Papers 11. Burton Historical Collection, Detroit Public Library.

———. 1895. *Indian Affairs 1801–1848*. Washington, DC.5: U.S. GPO.

Sagard, G. 1623. *The Jesuit Relations and Allies Documents*. Toronto: McClelland and Stewart Publishing.

Stothers, David M. 1999. Late Woodlands Models for Cultural Development in Southern Michigan. In *Retrieving Michigan's Buried Past: The Archaeology of the Great*

Lakes State, ed. J. A. Halsey. Bulletin 64. Bloomfield Hills, MI: Cranbrook Institute of Science. 194–210.

Walker, A. 1816. *A Journal of Two Campaigns of the Fourth Regiment of U.S. Infantry in the Michigan and Indiana Territories, under the Command of Col. John P. Boyd and Lt. Col. James Miller during the Years 1811–1813.* Keene, NH: Sentinel Press.

Zeisler, K. 1969. *A Brief History of Monroe.* Monroe, MI: Monroe Evening News. 1–8.

Zurel, R. 1961. *The Tribes between the Lakes.* Bloomfield Hills, MI: Cranbrook Institute of Science.

THE DETROIT RIVER AS AN ARTERY
OF TRADE AND COMMERCE

John K. Kerr, W. Steven Olinek, and John H. Hartig

In the 1600s the French were interested in expanding their empire into the New World. They had grandiose ideas for a New France and sent explorers to find a natural waterway up the St. Lawrence River, through the Great Lakes, across to the Mississippi River, and then south to New Orleans. They hoped this crescent-shaped transportation route would become the central avenue of communication for New France (Hatcher 1945).

In 1679, a thirty-five-year-old French explorer named La Salle sailed a sixty-foot-long vessel named the *Griffin* across Lake Erie and up the Detroit River. It was the first sail of a European across Lake Erie. As was customary during that time, a priest was onboard to minister to La Salle and his crew. That priest was Father Louis Hennepin, who also recorded the observations and sentiments of all onboard. As the *Griffin* sailed up the Detroit River and through Lake St. Clair in 1679, Father Hennepin wrote the following:

> This straight is finer than that of Niagara, being 30 leagues long, and everywhere one league broad, except in the middle which is wider, forming the Lake we have named St. Clair. The navigation is easy on both sides, the coast being low and even. It runs directly from north to south. The country between these two lakes is very well situated and the soil very fertile. The banks of the straight are vast meadows, and the prospect is terminated with some hills covered with vineyards, trees bearing good fruit, groves and forests so well disposed that one would think nature alone could not have made, without the help of art so charming a prospect. The country is stocked with stags, wild goats, and bears which are good for food, and not fierce as in other countries. (Hatcher 1945, 33–34)

La Salle's crew hunted along the shores of the Detroit River and filled the ship's galley with game. Wild turkeys were found in the forests of oak, chestnut, and walnut. Flocks of swans were a common site as the *Griffin* headed upriver. This voyage was exceptional because it laid the navigational foundation for European settlement of this area.

Most North American cities were founded primarily as outposts to stimulate trade and commerce for the mother country. Beginning in the 1600s, seaboard cities were considered nodes of a feeder system that provided raw materials for the homeland and a limited market for manufactured goods. Key factors in establishing these settlements were navigational access (i.e., good harbors); an ample supply of timber for ship stores; and rivers that drained potential hinterlands. This was also the case for interior cities like Detroit and St. Louis.

A young French officer named Antoine de la Mothe Cadillac established the first European settlement in Detroit in 1701. Detroit was established to expand trade and commerce, but the actual physical location was dictated for military reasons. Cadillac's vision was for a strong French post on the Detroit River that would be a bulwark against the British traders, a center of influence among the Indians, and a key bastion in the chain of forts from Quebec to the lower Mississippi (Hatcher 1945). During this era Detroit was called "The City of the Straights." In French the word *detroit* means narrows or straight. Cadillac's goal was to establish a military presence at the narrows to control water traffic and the fur trade that moved over the water. This was essential not only to advance the fur trade but also to preserve it (Johnson 1971). Not only did Detroit become a major center for collecting and exporting furs, it would become a major center for processing of furs.

Cadillac's plan for Detroit was to establish a full-service, self-supporting community providing the skills and services necessary to support the fur trade. Cadillac not only brought fruit trees, seeds, and domestic animals with him to Detroit, but he constructed a gristmill, brewery, icehouse, warehouse, fort (i.e., Fort Pontchartrain), lodging house, and church (Holli 1976). His community of French farmers, fur traders, soldiers, slaves, and Native Americans formed a French culture that dominated Detroit for the next 125 years.

Waterborne Commerce

Cadillac recognized the importance of the Detroit River as a means of communication, commerce, and trade. In its early years, Detroit's waterfront was primarily a natural shoreline with no major human-made structures until people ventured out of the fort to use canoes for transportation. In 1796 the first wharf was constructed in Detroit, called the Merchant's and King's Wharf (Levanen 2000). After the English surrender of Detroit to the United States in 1796, communication increased between the two sides of the Detroit River. Demand for ferry service increased and it became very profitable and subsequently had to be regulated. In 1802 the first license for a ferry service across the river was granted to Gabriel Godefoy to transport livestock, produce, and people (McLean 1994). In 1806, a ferry house was built on the Detroit waterfront approximately fifty feet west of Woodward Avenue between Atwater and Woodbridge Streets.

As waterborne commerce increased, the physical configuration Detroit's waterfront changed substantially. For example, Levanen (2000) reports that in 1816 a wharf ten feet wide and extending two hundred feet into the Detroit River was authorized. Fees were to be fifty cents to tie a boat five tons or under, coming a distance of ten miles or more. Boats up to twenty-five tons were to pay a dollar and a half, and larger craft had to pay two dollars.

In 1818 the first steamboat on the upper Great Lakes, *Walk-In-The-Water,* arrived in Detroit. Most of the city's population turned out for the arrival of this one-hundred-passenger steamboat. One cannot stress enough the economic and social importance of this first visit by *Walk-In-The-Water.* Everyone's life was changed. New settlers were encouraged to come to Detroit. The dreaded two-week voyage across Lake Erie was cut down to just a comfortable few days and a regular cruise schedule was established between Buffalo and Detroit.

With improved transportation it did not take long for the population of Detroit to increase. In 1817 it was 900. In 1830 it grew to 2,200 and by 1837 it had grown to 9,700. The Detroit River became so busy with sailing and steam vessels that Congress declared it a "public highway" on December 31, 1819 (Levanen 2000).

With the completion of the Erie Canal in 1825, the advantages of waterborne commerce increased dramatically. For example, overland travel had been hard, hazardous, and for commerce quite expensive. The cost of transporting a barrel of goods to Buffalo decreased from five dollars in 1815 to fifty cents in 1825 because of the efficiencies of water transportation (Levanen 2000). Commerce grew in volume each year. As Hatcher (1945) notes, "Detroit, as both a way station and an embarkation point for the lands farther west, felt the tremendous stirrings of the continent" (181).

In 1827 the Detroit City Council voted to improve the waterfront facilities with a sixty-foot-wide dock at the foot of Woodward Avenue. By 1857 shipments were coming into Detroit from Europe, and the port was becoming recognized internationally. By the late 1800s five miles of Detroit River waterfront were lined with docks (Levanen 2000).

Between 1850 and 1870 Detroit reached a high plateau of commercial development where out-of-state business with agricultural settlements had replaced trafficking in furs as the main line of business (see table 3.1). It was predominantly a commercial city. Well before Detroit became an industrial superpower and the Motor City it was a leading center of commerce. Holli (1976) concludes that Detroit's primary processing industries and trade-serving manufactures during 1850–70 were the building blocks and provided the knowledge for a factory town that mass-produced automobiles.

During 1850–70 wholesale trading and retailing were the two most prominent and economically vital functions of a mercantile city like Detroit. Roberts (1855) provides a description of a typical business establishment during this era:

The store has a free stone front, is four stories high, occupies a front of fifty feet, and extending in depth one hundred feet, comprising ten rooms, each twenty-five feet in width and one hundred feet in depth and an area of 25,000 square

TABLE 3.1
Principal Industries of Detroit, 1860

Industry	Value of Products ($)
Copper smelting	1,500,000
Lumber sawed	619,049
Machinery: steam engines, etc.	608,478
Iron: bar and railroad	585,000
Leather	380,225
Flour and meal	313,837
Liquors, malt	262,163
Iron, pig	145,000
Furs	143,000
Soaps and candles	137,915
Printing: newspaper and job	136,400
Boots and shoes	131,852
Sash, doors, and blinds	126,929
Bread and crackers	99,200

Source: Holli, 1976.

feet, all of which are filled to their utmost capacity with foreign and domestic dry goods, carpets, cloths, millinery and clothing—in addition to which the firm occupies a store-house in the rear. The retail rooms are four in number, finished in the most gorgeous style. About three hundred gas lights are required to light the several apartments. From sixty to seventy-five salesmen and from one hundred to one hundred and fifty persons altogether are employed in the several departments, and including those outside, seamsters and seamstresses, the firm gives employment to about six hundred persons. Their invoices of merchandise imported during the year 1854 amounted to more than seven hundred thousand dollars. This store was recently refitted and opened for the fall trade with an invoice of goods amounting to over four hundred thousand dollars.

Another important export commodity during this time was fish. Detroit River whitefish was often called the "delicacy of delicacies." Holli (1976) notes that it was not uncommon to see Detroit River whitefish advertised in St. Louis, Cincinnati, Louisville, and Natchez. It was estimated that in 1855 the annual whitefish catch was 100,000 barrels. A commercial fishery in the Detroit River first developed in the early 1800s, but catch records reflect closure of the commercial fishery in 1909 in Michigan waters to promote sport fishing (Manny, Edsall, and Jaworski 1988). An Ontario commercial fishery continued through 1970 when it was closed due to the mercury crisis (see chapter 8). It is believed that the abundance of whitefish in the Detroit River declined due to overfishing, pollution, and loss of habitat.

Detroit's population increased from about 45,000 in 1860 to 285,704 in 1900 (Dunbar 1965). Much of this growth can be attributed to the development of manufacturing industries. Dunbar (1965) describes the reasons for expansion of manufacturing in the following account:

> The easy availability of iron, copper, lead, wood, and other raw materials had much to do with the location of important manufactures. Boxes and barrels for shipping were cheap at Detroit; coal had to be shipped a relatively short distance; railroad and water transport facilities were excellent. By 1880 it was estimated that almost $16 million was invested in manufacturing plants, and that the annual product was worth $35 million. Iron and steel industries were the most important Detroit industries in the 1880s. The largest factory for the manufacture of railroad cars and car wheels in the nation was located at Detroit. Immense stove factories, notably the Detroit Stove Company, the Peninsular Stove Company, and the Michigan Stove Company, were front-rank industries. Parke, Davis and Company, famed maker of pharmeceuticals, had its beginning in Detroit in 1867. In 1880 over sixty establishments manufactured chewing tobacco and cigars, and the city was one of the largest manufacturers of chewing tobacco in

the country. The show factory in which Hazen S. Pingree made his fortune was the largest of its kind west of New York. D. M. Ferry and Company was the leading concern in the growing and distribution of seeds. The 919 manufacturing establishments in Detroit in 1880 employed some 16,000 persons.

Waterborne commerce increased substantially during this era in direct response to the growth of commerce (figure 3.1). In 1850 a total of 2,341 vessels carrying a total of 671,545 tons passed through the port of Detroit; in 1907 a total of 75,000,000 tons passed through. Although tonnage figures have dropped in recent years because other forms of transportation have become available and due to commodity demands, the average tonnage to pass through the port of Detroit has hovered near the 30,000,000 mark (Levanen 2000).

One cannot talk about the port of Detroit without referencing the St. Lawrence Seaway. Said to be the "engineering marvel of the century," the seaway and its fifteen locks represent more than that to the city and port of Detroit. It is also referred to as "the gateway to the global marketplace" or "gateway to America's heartland." The seaway has brought jobs, cargo for local manufacturing, and localized economic impact that has helped fuel the

Figure 3.1. The Detroit River waterfront and view of the city of Detroit in 1889 (Calvet Lithographic Company, Detroit; Library of Congress Panoramic Maps, 2nd ed., 342).

nation's automobile, grain, and construction industries to name a few. Created through one of the most intense, large-scale construction projects, the seaway promises to remain a permanent gateway to the world. The port of Detroit is at its heart.

Before the French settled in Detroit in 1701, the famous French explorer Jacques Cartier was busy exploring the canals in and around the St. Lawrence River basin. Cartier soon recognized that he would not be able to trek further into the river due to the rushing waters caused by waterfalls and sea level differences. Over the next several decades, links between Lake Erie and the Atlantic Ocean began to develop.

Meanwhile, with the French establishment of Detroit, important and reliable access was established between the upper and lower Great Lakes. Forest, fertile soil, and mining materials were abundant and helped create a foundation for Detroit's reputation as a viable and rich manufacturing center.

Through the 1800s, the port of Detroit remained a vibrant port of call and a marketplace for domestic goods. Products traded and sold included various mining commodities including copper as well as lumber and aggregates. Additionally, the ties first forged in the late 1700s began to affect Detroit with the creation of a canal between the Great Lakes and Hudson River (Plumb 1911). Another significant development was the opening of the Erie Canal. Through these events, Detroit soon became a gateway to the unexplored lands of the Northwest Territory.

Detroit was rapidly developing as a center for trade. Its logistical importance as a vital link to key commodities, necessary to feed the growing nation's economy, became nationally recognizable. With timber, wool, and agricultural products available to the north and west and manufactured goods available from the East Coast, logical trade relationships formed. Added to that were crucial supplies of limestone, copper, and iron ore from northern Michigan and coal from the south. Soon Detroit became home to one of the busiest ports in the world.

A variety of services and jobs were abundant throughout the port of Detroit. Dock laborers and shipbuilders could find employment, and bankers, importers, and storage companies found that their services were increasingly needed. The Detroit River and port of Detroit helped establish a full-scale and diversified economy for the city and provided many with the hope and ability to start a new life within the rapidly growing region.

The first lock at Sault Ste. Marie was built in 1855 to help facilitate this trade with the northern territories (Cuthbertson 1931). In 1881, the state of Michigan donated this lock to the U.S. government. Through the early part of the 1900s two more locks were built, the Davis and Sabin locks, providing access for a growing fleet and mining establishments to the north.

Minnesota Congressman John Lind is credited with thinking up the "seaway idea" at the turn of the twentieth century. His idea was to create a deepwater route from the head of Lake Superior, and its ports in Duluth, to the Atlantic Ocean and ports of call in Montreal and Quebec. His resolution to the Minnesota legislature was passed in 1895 (The Saint Lawrence Seaway Development Corporation 2001).

With pressure mounting within the federal government to address this issue, and with simultaneous Canadian interest beginning to play a role, legislative action was soon introduced. In 1909, the Boundary Waters Treaty was created, establishing the International Joint Commission (IJC), a regulatory agency that would monitor development along the international waterway, the St. Lawrence Seaway. In 1921, the IJC recommended that improvements to the St. Lawrence River become a priority in order to accommodate burgeoning trade between the inner Great Lakes ports and Canada, as well as other global trading partners.

There was great debate in the federal legislature before this idea became a reality. Most of the objection came from the coal and rail industries, as both stood to have business diverted if such a system came into being. For the next several decades, they succeeded in blocking development of a full-scale St. Lawrence Seaway/Great Lakes maritime transportation system.

However, during this time Canadian interest in developing this trade corridor grew. With the United States reluctant to pledge funding for the research and development of such a connection, Canada began to do so on its own. Taking this into account, a new Congress under newly elected President Dwight Eisenhower passed legislation that would allow the United States to split the construction of the St. Lawrence Seaway with Canada.

In the early 1950s, iron ore was discovered in Canada's Labrador wilderness. This discovery would eventually tip the scales in the great debate toward building an international seaway, as iron ore was used in large part by the domestic steel industry to create the slabs and plates for the booming automobile industry. It became apparent that development of the seaway needed to begin; otherwise the United States risked being left behind in the mostly subtle international rivalry between Canada and the United States (and especially Detroit) for the rich mined product to the north.

Sponsored in part by George Dondero (R-MI), the Wiley-Dondero Act passed in 1954 and allowed the United States to construct, operate, maintain, and develop the seaway. This event marked the birth of Detroit as a modern world port of call and allowed the many international vessels into our waters that still call on our port today and deliver over two million tons of foreign products.

Shipbuilding

Because of the Detroit area's strategic location along the Great Lakes and its position as a center of commerce, as well as the demand for transportation of passengers and freight and the availability of essential resources, the Detroit area became one of the greatest shipbuilding ports in the United States. The British established the first shipyard along the Detroit River in 1760 to produce armed naval vessels and commercial sailing craft. Levanen (2000) provides the following account of the history of shipbuilding in Detroit:

> Hundreds of ships were built along the waterfront. In 1907, twenty-one of the largest ships sailing the lakes were launched in Detroit. The waterfront was dotted with shipbuilding rigs, launching docks, and dry docks. To aid the war effort in both World Wars Detroit yards turned out military vessels. . . . One third of all marine engines built in the United States from 1897 to 1920 were built in Detroit. Some of these famous builders were Riverside Iron Works, located at the foot of Chene Street, Frontier Iron Works, situated near Belle Isle, and the Dry Dock Engineering Works, located at the foot of Orleans Street. (9)

It should be no surprise that as a result of shipbuilding, Detroit became a major center for production of paint varnish, steam and gasoline engines, metal pipe and parts, and over one hundred other marine parts. In the 1890s more ships were built in the Detroit area along the Detroit River than any other city in America. It is believed that the knowledge gained from building steam engines for ships laid the foundation for automobile manufacturing.

Shipbuilding also had a significant impact on communities along the lower end of the Detroit River. From 1801 to 1960 at least 655 ships were built at yards and locations from the lower Rouge River downstream to the mouth of the Detroit River (figure 3.2). Again, the significance of the Detroit River as a shipbuilding center and transportation corridor was manifested by the fact that on December 3, 1819, the Detroit River was declared a public highway by an act of Congress. In addition, more passenger trade went out of Detroit than anywhere in the world during the 1890s. Shipbuilding provided jobs and supported local families. It also had a significant

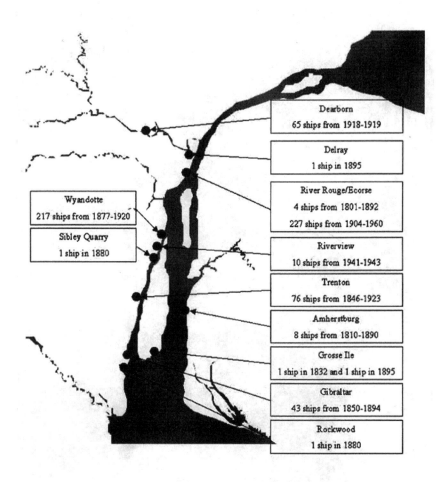

Figure 3.2. An overview of shipbuilding history along the lower end of the Detroit River.

impact on our region, the nation, and the world. Over the years these ships would help further billions of dollars of commerce and trade.

Cross-border Trade and Commerce

It was not until the 1890s that heavy industrial manufacturing became Detroit's most important business. During that era, hundreds of factories, forges, foundries, and dry docks transformed the raw materials of iron, copper, and wood into finished products. The most important products were railroad cars, ships, and stoves. In fact, Detroit was the leading producer of stoves in the 1890s. Thousands of skilled and unskilled Detroiters were employed in these industries. Railroad car manufacturing and stove making had the largest and second largest industrial workforces in Detroit, respectively, followed by stove making (Detroit Historical Museums 2001). The factories that lined the riverfront and city railways brought wealth and great prosperity to the region.

Long before the start of automobile manufacturing, Detroit was a center for a wide variety of industries. The region provided plentiful resources, the Detroit River provided efficient transportation, and Detroit and the Midwest provided a growing market. Timber from the region was used to manufacture furniture and railroad cars. Copper and iron ore brought in by ship made Detroit the largest producer of stoves, furnaces, and heating radiators. The salt deposits lying beneath the Detroit River attracted J. B. Ford to build a chemical plant in Wyandotte in 1891. It later became Michigan Aklali, then Wyandotte Chemical, and now BASF Corporation.

As noted earlier, the first ferry service across the Detroit River was established in 1802. In fact, Windsor became a major Canadian riverfront community because it was the site of the ferry link to Detroit and later the Great Western Railway terminus, which had been attracted by the ferry crossing (Essex Region Conservation Authority 1999). The first railroad car ferry to cross the Detroit River was the steamer *Great Western,* launched in 1867. In 1910 the Michigan Central Railway Tunnel opened and transported 243,000 railroad cars (McLean 1994). Interest in more rapid and versatile transportation led to the construction of the Ambassador Bridge between 1925 and 1936. The Detroit-Windsor Tunnel opened for vehicular traffic in 1930. Ever since their construction, the bridge and tunnel have been of enormous importance to commerce and industry and have strengthened the bond between two great nations. The inscription etched in identical tablets at each end of the Ambassador Bridge reads: "The visible expression of friendship in the hearts of two peoples with like ideas and ideals."

The beginning of the twentieth century marked another turning point in the history and development of the Detroit area. By 1900 Detroit's commercial advantage had vanished. Detroit had become an industrial city. Automobile manufacturing would soon dominate the economy of the Detroit area. Holli (1976) reports that in 1904, 3.8 percent of Detroit's 60,554 industrial employees were employed in the automobile industry; in 1919, 45 percent of Detroit's 308,520 industrial employees were employed in the automobile industry. Detroit was now the Motor City and one of the largest industrial manufacturing centers in the world.

Today, economic trade and commerce across the Detroit River are significant. For example, the United States and Canada form the world's largest trading relationship (Embassy of Canada 2000). During 1998, total U.S.–Canadian trade exceeded $396 billion. Approximately one-third of all trade that travels by road passes over the Ambassador Bridge. In addition, approximately 22 million vehicles passed through the Detroit-Windsor Tunnel or over the Ambassador Bridge during 1998.

The automotive industries of Michigan and Canada are highly integrated. In 2000, they exchanged $36.8 billion in automobiles and trucks, and $16.3 billion in motor vehicle engines and other parts, most of this across the Ambassador Bridge (Embassy of Canada 2000).

Rumrunner Era

It would be an oversight to not mention the "rumrunners." In 1920, the Eighteenth Amendment and the Volstead Act prohibited the production and sale of alcohol in the United States. Rumrunners immediately began smuggling beer and liquor from Canada to the United States across the Detroit River. In fact, an estimated 75 percent of all liquor smuggled into midwestern states during prohibition came in through the tunnel. Because of the relative short distance across the river, the number of islands, and the number of small docks and landing places, the river was an ideal place to smuggle alcohol.

Detroit Historical Museums (2001) reports that in one year more than 3,000,000 gallons of Canadian beer and whiskey were exported to the United States, yet authorities intercepted only 5 percent of the liquor. Mason (1995) notes that the Detroit Police Department had only one patrol boat in 1920, which was a "good seaworthy scow that by an effort could overhaul a tugboat" (118–19). Rumrunners were "running circles around us," the police commissioner at the time complained. During Prohibition liquor of all varieties that came across the Detroit River was readily available throughout Detroit and the rest of Michigan.

Unintended Consequences

The Detroit River played a huge role in the history and development of the Detroit-Windsor metropolitan area. It literally served as an artery for commercial development and economic expansion. This produced numerous benefits to the region. However, with the expansion of trade, commerce, industry, and population would come a number of unintended natural resource and environmental problems that are discussed in subsequent chapters. Increased population and greater affluence begot effluents and resource degradation. We must not forget these problems and use them as lessons for ensuring environmentally sustainable economic development.

Literature Cited

Cuthbertson, G. A. 1931. *Freshwater: A History and Narrative of the Great Lakes.* Toronto: MacMillan.

Detroit Historical Museums. 2001. *Virtual Tours: Glimpses of Michigan's Past.* http://www.detroithistorical.org.

Dunbar, W. F. 1965. *Michigan: A History of the Wolverine State.* Grand Rapids, MI: W. B. Eerdmans Pub. Co.

Embassy of Canada. 2000. *United States-Canada: The World's Largest Trading Relationship.* Washington, DC: Foreign Affairs and International Trade.

Essex Region Conservation Authority. 1999. Detroit River Nomination Document—Canadian Heritage Rivers System. Essex, Ont.

Hatcher, H. 1945. *Lake Erie.* New York: Bobbs-Merrill.

Holli, M. G., ed. 1976. *Detroit.* New York: New Viewpoints.

Johnson, I. A. 1971. *The Michigan Fur Trade.* Lansing, MI: Black Letter Press.

Levanen, A. J. 2000. "The River's Edge: A History of the Detroit Waterfront." *Telescope* XLVIII:2 (March–April): 6–9.

Manny, B. A., T. A. Edsall, and E. Jaworski. 1988. *The Detroit River, Michigan: An Ecological Profile.* Biological Report 85(7.3). Ann Arbor, MI: U.S. Fish and Wildlife Service.

Mason, P. P. 1995. "The Prohibition Navy: Enforcement of the Volstead Act." *Telescope* XLIII:5 (September–October): 117–24.

McLean, E. G. 1994. "Detroit River Crossing." *Telescope* XLII:3 (May–June): 59–64.

Plumb, R. G. 1911. *History of the Navigation of the Great Lakes.* Washington, DC: U.S. GPO.

Roberts, R. E. 1855. *Detroit: Sketches of the City of Detroit, 1855.* Detroit, MI: Saint Lawrence Seaway Development Corporation

The Saint Lawrence Seaway Development Corporation. 2001. *A History of the Great Lakes Seaway System. http://www.seaway.dot.gov.*

Zacharias, P. 2000. *The Building of the Ambassador Bridge. http://www.detnews.com.*

✦ 4 ✦

AMERICAN BEAVER EXPLOITATION FOR EUROPEAN CHIC

John H. Hartig

Native Americans first lived along the banks of the Detroit River as hunters and gatherers. Supplies of fish, deer, other animals, and freshwater were the primary reasons for earliest Native American settlements along the Detroit River. These First Nations also gathered fruits, grains, and nuts to meet their nutritional needs. With Native American population densities relatively low, it was easy to meet nutritional needs without overexploiting resources.

However, as the Native American population grew in the watershed surrounding the Detroit River and European settlement began, a society dominated by hunting and gathering shifted to one incorporating farming and some trading. To live as a hunting and gathering society requires considerable land. Too many people were trying to live off of too little land. This shift was typical throughout North America as human populations grew and natural resource limitations became manifest (Muir 2000).

One of the first North American examples of overexploitation of natural resources by humans was the American beaver. Native people evolved from hunters and gatherers to living by hunting, fishing, farming, and trading. Furs became important trade goods, especially after the 1600s when Europeans established outposts on the Great Lakes. Native people exchanged pelts of beaver and other animals for European-made knives, tools, blankets, iron pots, rifles, and other goods. But what was driving this demand for beaver pelts?

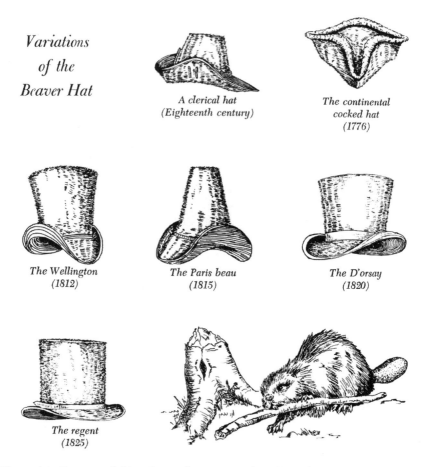

Variations of the Beaver Hat

A clerical hat
(Eighteenth century)

The continental
cocked hat
(1776)

The Wellington
(1812)

The Paris beau
(1815)

The D'orsay
(1820)

The regent
(1825)

Figure 4.1. European fashion during the seventeenth, eighteenth, and nineteenth centuries demanded fur hats made of beaver pelts (North American Fur Trade Conference 1967).

European Demand for Beaver Pelts

During the late 1600s and early 1700s in Europe, big hats had gone out of style and beaver pelt hats had come into style. European fashion during that era demanded fur hats made from beaver pelts (figure 4.1). Muir (2000) has described this millinery shift as follows:

> The soft hats of the Renaissance—those large, velvet berets of gathered brim and elegant plume familiar from Hobein's portrait of Henry VIII—went out of style,

and hats made of felt came in. These were the wide-brimmed, round-crowned hat of the Quaker merchant and the rakishly upswept and befeathered chapeau of the Restoration cavalier, the plumed hats of the Three Musketeers and the sober, high-crowned buckled hats of the Pilgrim fathers. Sober or extravagant, pious or worldly, modest shopkeeper or wealthy merchant: for nearly three centuries every European man who could afford a hat at all had one made of felt. Very often the women did too. (26)

As these hats grew in popularity, demand for European beaver pelts grew. Muir (2000) notes, "Beaver, once as common in the Old World as the New, had been eradicated everywhere in Europe except for eastern Russia and the steep valleys of the Pyrenees, a region far too small to supply the fashion needs of a prosperous continent" (26). As overexploitation of beaver occurred in Europe to meet a growing mercantile attraction, explorers began to look at North America as a potential source of beaver.

Early European explorers first came to North America in search of the spice-rich Orient but were disappointed. North America did, however, have a beaver population numbering in the millions that could compensate for the dwindling beaver population in Europe and help meet the mercantile demand in Europe.

Beaver Natural History and Exploitation

Beavers are the largest rodents in North America. They are relatively short (adults grow to 2–3 feet or 61–91 cm) and squat, with a large rump. Each foot has five toes, and the hind feet are webbed to help with swimming. The second toe on each hind foot has an extra claw (a double claw) to help the beaver groom. Typically the flattened tail is used to help balance and maneuver, but it is also used to issue a warning signal when slapped against the water.

Beavers have two different types of hair: the guard hair, which is up to two inches in length, and the "underhair" or fur, which is a maximum of one inch long. Guard hair is long tough hair that grows through a shorter softer layer of hair. This underhair helps insulate the beaver in cold or wet weather. Beavers' guard hair is a dark chestnut brown, while the underhair is usually a softer, reddish-brown. The beaver pelts that commanded the highest prices were those that had been worn or slept in by the Native Americans and were known as *castor gras* or greasy beaver (Dunbar and May 1995). Body sweat from Native Americans and smoke from their lodges made these used pelts supple and loosened the long, coarse guard hair, which was then easily removed, leaving the soft, high-quality underhair. An average adult beaver weighs about 30–60 pounds (13.6–27.3 kg) and the pelt from such an animal might weigh 1.5–2 pounds (0.7–0.9 kg).

The two incisors of a beaver's jaw grow continuously as they are worn away. Their chewing force is 176 pounds (80 kg), compared to human's chewing force of 88 pounds (40 kg). Despite having large front teeth, these animals are herbivores. Most commonly they eat the bark of certain hardwoods such as poplar, aspen, birch, cherry, willow, maple, and alder. Aspen and pondweeds are their favorite natural foods.

Beavers are almost entirely aquatic, although they travel on land to reach nearby trees. They are well-known for their engineering feats. A single pond or lake will consist of a family group living in burrows in the shoreline or beaver lodges built of mud, stones, and tree branches. Access to both burrows and lodges is subsurface for protection from predators. Due to the construction of the lodge, many predators are unwilling to take the time to dig through the walls to get at the beavers. To assure adequate water depth, streams are dammed using logs, branches, stones, and mud. Lodges are usually 8–12 feet (2.4–3.7 m) in diameter and have heights of around 3 feet (0.9 m), but are sometimes larger.

Beavers are very efficient swimmers, using their hind feet and tail. They are physiologically adapted to aquatic life and can stay underwater for up to fifteen minutes. Valves in the nose and ears close automatically upon submerging; their mouth closes behind the incisors. The eyes have membranes that can be drawn over the eyeball. They have an oversized liver to deal rapidly with byproducts, a large lung capacity, and high tolerance to carbon dioxide. Their efficient digestive system allows for utilization of a large cellulose intake.

Beavers tend to be monogamous. Family groups typically consist of two adults, several two-year-olds, and the young of the current year. They first mate at about three years old. The gestation period is 128 days. Two to six kits per litter are born in the spring. They can swim when a few hours old and are weaned at one month. Mothers can carry kits in their mouth, with support from her front legs while walking upright on her hind legs and tail. Young leave or are forced out of the colony at about two years old. Beavers are social animals. Family life is cooperative and all members help with the incessant hard work of food gathering and lodge building.

Beaver became an easy mark because they do not migrate, live in lodges that can be attacked at any time in all seasons, and increase in population size by only about 20 percent annually. Thus, it is easy to see why the beaver was exterminated in one region after another and traders were forced to move further and further inland to find them.

Prior to the arrival of Europeans, beaver hunting occurred but did not diminish the population. Native Americans hunted with primitive weapons. With European arrival and use of iron weapons and guns, beaver were soon slaughtered and virtually disappeared from the area. It is well accepted that two factors contributed to the decline of the beaver population: overhunting and clearing of the wilderness for farmland and settlements, which resulted in substantial loss of beaver habitat.

The beaver was this continent's first natural resource to be overexploited. In fact, at one time, beaver pelts were the unit of currency in the new land. Native Americans knew how to hunt and, with beaver skins, they knew they could purchase whatever they needed from the European traders. The beaver was so common in Detroit that native people referred to the area as "the country of the beaver" (Rosalita 2000, 3). French settlers later named Beaver Street in Detroit because of the important role this animal played in early history.

It is estimated that there were about ten million beaver in America when the Europeans arrived (Dunbar and May 1995). Europeans exported 50,000 skins annually until, by 1800, numbers of beaver were substantially reduced east of the Mississippi. During the height of beaver exploitation during the eighteenth and nineteenth centuries, hundreds of thousands of beaver skins were exported to Europe from North America annually. These animals were also sometimes destroyed because of the damage they did to forests and the flooding occasionally caused by their damming practices.

By the mid-1800s the fur trade in Detroit was over. Detroit merchants no longer depended on furs. Commerce, agriculture, and lumbering became the basis of the next economy. Ceaseless slaughter nearly wiped out the beaver populations in Europe and North America east of the Mississippi River by 1930. The beaver is still almost extinct in Europe but is becoming reestablished in Canada and in protected areas of the United States, primarily due to conservation measures and a fashion shift to silk hats. Danger of extinction now seems remote. In some places the beaver has now reestablished itself to nuisance proportions.

It is worth noting that in Canada, the trade of beaver pelts became so lucrative that the Hudson Bay Company honored this buck-toothed rodent by putting it on the shield of its coat of arms in 1678. The Hudson Bay Company shield consists of four beavers separated by a red St. George's

Cross and reflects the importance of this animal to the company. A Canadian coin was also created to equal the value of one beaver pelt. In 1975 the beaver became an official emblem and symbol of the sovereignty of Canada.

Tragedy of the Commons

The story of beaver exploitation is an important common-property resource lesson. Quite frequently a dilemma occurs in the use of common-property resources, which are owned by no one and available for use by everyone. It is difficult to exclude anyone from using them and each user depletes or degrades the available supply of these resources.

When the number of users of common-property resources is relatively small, there is no problem. However, over time the cumulative effect of many people trying to maximize their use of the resources impacts the available supply of resources. This abuse or depletion of these common-property resources becomes a "tragedy of the commons" when eventually no one can make a profit or otherwise benefit from the resources (Hardin 1968).

The "tragedy of the commons" is frequently told through a story of raising cattle on an open pasture or commons. As long as there is sufficient pasture land, shared real estate (i.e., commons) is efficient and no fences need be maintained and no human supervision is required (Hardin 1993). Such pasture lands are called "unmanaged commons." Ecologists are quick to point out that such an arrangement will only work if the number of herdsmen and cattle is kept below the carrying capacity. The carrying capacity is the maximum number of cattle that the pasture can support over a given period of time. However, if the number of cattle in the pasture exceeds the carrying capacity, the amount of useable pasture is depleted or degraded. Put another way, as the number of people and cattle increases, the amount of pasture land becomes a limiting factor to animal production, either milk to drink or beef to eat.

However, each pasture is of a given or finite size. Can each herdsman keep as many cattle in the pasture as he wants? Eventually each herdsman asks either implicitly or explicitly, "What if I added one more cow to my herd?" The benefit would be that the herdsman would get the proceeds from the sale of one more cow. The negative consequence might be potential overgrazing created by one more cow. The benefits of adding one more cow would be accrued by the herdsman; however, all herdsmen would share the negative consequences. Each herdsman eventually concludes that the only sensible course of action is for him to add one more cow to his herd. Then he decides to add another, and another, and another. Every herdsman sharing the pasture reaches this same conclusion. Therefore, each herdsman is compelled to increase his herd without limit in a pasture that has limited carrying capacity. Each herdsman pursues his own self-interest in a pasture with limited carrying capacity. Freedom of choice in the pasture brings depletion and degradation of the available grazing land in the pasture (Hardin 1968).

This "tragedy of the commons" is precisely what happened to the American beaver population in Europe, North America, and specifically the Detroit River watershed. Before Europeans came to our region, Native Americans hunted beaver but not to the extent that they depleted the population. Native American population density was relatively low, and Native Americans did not overhunt the beaver. Then Europeans came for beaver.

In the 1600s the French were interested in expanding their empire into the New World. Detroit was viewed as an extension of Montreal and the French maritime settlements of the North Atlantic coast (Holli 1976). The French wanted to capture the fur trade and whatever other riches might be unearthed there.

The first French settlement in Detroit was established 1701 by a young officer named Antoine de la Mothe Cadillac. Cadillac praised the region along the strait in his first report to Count Pontchartrain (Bald 1954):

> The banks [of the river] are so many vast meadows where the freshness of these beautiful streams keeps the grass always green. These same meadows are fringed with long and broad avenues of fruit trees which have never felt the careful hand of the watchful gardener; and fruit trees, young and old, droop under the weight and multitude of their fruit, and bend their branches toward the fertile soil which has produced them. . . . [The area is] so temperate, so fertile, and so beautiful that it may justly be called the earthly paradise of North America. (52)

Cadillac wanted a strong French post on the Detroit River that would be a bulwark against the British traders, a center of influence among the Indians, and a key bastion in the chain of forts from Quebec to the lower Mississippi (Hatcher 1945). Cadillac chose the highest bank on the narrowest part of the river to establish his military post. Holli (1976) notes the narrowest part of the river was determined to be a distance that the French measured by cannon as only "one gunshot" across.

Cadillac's goal was to establish a military presence at the narrows to control water traffic and the fur trade that moved over the water. This was

Figure 4.2. Detroit was a major center for the fur trade (painting titled *Furs—Trappers—Traders* by Robert Thom; ©1965, Michigan Bell Telephone Company, all rights reserved).

essential not only to advance the fur trade but to preserve it (Johnson 1971). Not only did Detroit become a major center for collecting and exporting furs, it soon became a major center for the processing of furs.

At first, Native Americans brought their beaver pelts to the Detroit post and exchanged them for French goods (figure 4.2). However, as the beaver population dwindled in the immediate area, authorities sent licensed fur traders into the interior to trade with more distant Native Americans in their villages (Dunbar and May 1995). However, fur trade was not confined to just licensed fur traders. Their counterpart was the unlicensed, lawless trader of the woods called the "coureur de bois." Johnson (1971) describes these picturesque figures of the pioneer fur trade as follows:

> Mostly they were French or half-breeds, who believed that the furred creatures of the forest were the monopoly of neither king nor company, but the spoils of all, and hence they defied all law to the contrary. They plunged into the wilderness to barter with the savage, adopted his customs and his life, and married his daughters. Men who could ply paddle, hunt, trap, and speak the Indian tongue almost as well as the savage himself; gay, lighthearted, care-free, capable of great

endurance, a strange mixture of civilization and savagery, with a predominance of the latter,—such were the rovers of the wood, the coureurs de bois of the fur trading regime. (155)

European demand for furs had a devastating impact on the beaver population. As noted earlier, Europeans exported 50,000 skins annually until, by 1800, numbers of beaver were substantially reduced east of the Mississippi. As beaver became scarce from overhunting and clearing of the wilderness for farmland, it eliminated the source of income that Native Americans had relied on for hundreds of years. As a result of this limited income, Native Americans were coerced into signing unfair treaties (Detroit Historical Museums 2000). History has shown that the fur traders, wanting to be repaid for outstanding debts, put pressure on the Native Americans to sign away their land in exchange for cash. Most of the money from treaties went directly into the pockets of the traders, and often large amounts of treaty lands were granted directly to the children that fur traders had with their Native American wives.

Muir (2000) suggests that Native Americans did not kill beaver in a conscious decision to exterminate them or deplete the population; they hunted because they were swept up economic forces too overwhelming to resist. The European demand for beaver and the return on beaver was just so great that the beaver was almost hunted to extinction. If a hunter came upon a beaver colony and did not slaughter every animal present, another hunter would soon finish the job. As Muir (2000) suggests: "The choice was not kill or conserve, it was take the wampum or let somebody else have it; a brutal crash course in market economics" (34).

Concluding Remarks

The story of beaver exploitation is not a call for restoring the beaver population in the Detroit River watershed. We probably cannot restore the beaver's habitat in the midst of the Detroit and Windsor metropolitan areas. However, the story of beaver exploitation is a call for living sustainably and managing the uses of earth's natural resources by not allowing them to be depleted or deteriorated.

There would be other examples of overexploitation of animals and loss of critical habitats, but the American beaver was the first example in our region. We must learn and often relearn that there are limits to the use of natural resources. Our challenge is to live within these limits and meet our own needs without compromising the ability of future generations to meet their needs. This story of American beaver exploitation for European chic and the stories to follow are important lessons to remember in our efforts to sustain our communities, economies, and our ecosystems. For as George Santayana (1863–1952) is often quoted: "Those who cannot remember the past are condemned to repeat it."

LITERATURE CITED

Bald, F. C. 1954. *Michigan in Four Centuries.* New York: Harper and Brothers.

Detroit Historical Museums. 2000. *Glimpses of Michigan's Past. http://www.detroithistor ical.org.*

Dunbar, W. F., and G. S. May. 1995. *Michigan: A History of the Wolverine State.* 3d ed. Grand Rapids, MI: Eerdmans.

Hardin, G. 1968. *The Tragedy of the Commons. Science* 162:1243–48.

———. 1993. *Living within Limits: Ecology, Economics, and Population Taboos.* New York: Oxford University Press.

Hatcher, H. 1945. *Lake Erie.* New York: Bobbs-Merrill.

Holli, M. G., ed. 1976. *Detroit.* New York: New Viewpoints.

Johnson, I. A. 1971. *The Michigan Fur Trade.* Lansing, MI: Black Letter Press.

Muir, D. 2000. *Reflections on Bullough's Pond: Economy and Ecosystem in New England.* Hanover, NH: University of New England Press.

North American Fur Trade Conference. 1967. Aspects of the Fur Trade. Paper published by the Conference. St. Paul, MN.

Rosalita, M. 2000. *Detroit: The Story of Street Names. http://www.detroithistorical.org.*

Waterborne Disease Epidemics during the 1800s and Early 1900s

John H. Hartig

When Cadillac first came ashore in 1701 in an area near what is now known as downtown Detroit, he made camp and ordered his men to proceed with clearing land at the top of a bluff for their settlement. Logs were trimmed and hewed for cabins, a church, and a stockade. The location of this European settlement was chosen to control and guard the primary transportation route at that time—the Detroit River.

During these early years of European settlement, the Detroit River provided pure, wholesome water to residents. Early colonists walked down to the river, dipped a bucket into the water, and retrieved enough water for their daily needs (Farmer 1890). Soon after, it became common for the water to be stored in large wooden jugs that were buried in the ground to keep it cool. One can imagine what it was like to drink water that had stood for twenty-four hours during the warm summer months in wooden jugs stored aboveground.

Although this Detroit settlement had humble beginnings, it would not remain a relatively small French post for long. In 1748, settlers were offered inducements to come to Detroit (Nolan 2000). These included a spade, an axe, a plough, a large wagon, a small wagon and seed, a cow, and a pig, to be returned by the third harvest. Long, narrow strip farms would later be established perpendicular to the Detroit River, many extending two miles back from the river. As the population grew, more efficient methods of acquiring water were sought.

By 1812, there were 800 people and 160 houses in Detroit. The

Detroit River continued to be a source of safe drinking water and a primary transportation route. *Walk-In-The-Water* was the first steamboat to land in Detroit in 1818. Steamboat service reduced travel time between Buffalo and Detroit from 10 days to 42 hours. As more and more settlers came, Detroit had to accommodate the needs of its growing population. In 1827 a water system franchise was granted in Detroit, allowing water to be pumped from the Detroit River to a reservoir and then transported out through pipes made of logs (Nolan 2000). This water service was provided to 1,500 citizens of Detroit at a cost of $10 per year.

In the 1830s a number of other private reservoirs were constructed throughout the city to store drinking water pumped from the Detroit River. The largest of these reservoirs was called the Old Round House and located at the foot of Orleans Street (Farmer 1890). It was completed in 1838 and had a capacity of 423,000 gallons.

In 1853, Detroit's city charter was amended to establish a five-member Board of Water Commissioners. This board was granted special powers and authority to operate Detroit's water system. Over time the board would oversee the growth and continuous improvement of Detroit's water system to meet the needs of a rapidly growing Detroit population. It would become one of the largest systems in the United States in terms of water produced and population served.

The First Reports of Water Pollution

With the continued growth and development of Detroit in the 1800s came a number of problems, including water pollution. The first major water pollution problems were lack of sanitation, contamination of the public water supply, and resultant waterborne disease epidemics.

Waterborne disease epidemics have been reported throughout civilization. In fact, there are references to deaths due to dehydrating diarrhea dating back to Hippocrates and Sanskrit writings. However, it must have been frightening during the 1800s and early 1900s to see people in Detroit dying from a waterborne disease for which there was no real treatment. People came down with the disease and died a few days later.

Cholera is an acute illness characterized by diarrhea and/or vomiting. It is caused by infection of the intestine with the bacterium *Vibrio cholerae.* The primary route of exposure is consumption of water or food contaminated with *Vibrio cholerae.* The infection is often mild or without symptoms, but sometimes it can be severe. Approximately one in twenty infected persons has severe symptoms characterized by profuse watery diarrhea, vomiting, and leg cramps. In these persons, rapid loss of body fluids leads to dehydration and shock. Without treatment, death can occur within hours. In the United States, cholera was prevalent in the 1800s but has been virtually eliminated by modern sewage and water treatment systems.

Cholera was believed to have been brought over on passenger ships from Europe in the early 1800s. May (1995) provides the following historical account:

> Just one year before, on July 4, 1832, the steamer *Henry Clay* arrived at Detroit with several companies of soldiers under the command of General Winfield Scott, en route to meet the threat of Black Hawk's warriors. Several cases of Asiatic cholera had broken out, and when the ship stopped at Detroit, the extremely contagious disease was spread to the townspeople. It had appeared in Russia in 1831, had spread to Western Europe, and was brought to America by a passenger on an emigrant ship. It soon made its appearance in cities on the eastern seaboard, where many deaths were reported. At Detroit there were fifty-eight cases, and twenty-eight deaths took place within two weeks. The territorial capitol was utilized as a hospital for the victims, and the townspeople were in a state of panic. Out state towns set up guards to prevent travelers from Detroit from using the public roads. A stagecoach with passengers, attempting to elude the roadblock east of Ypsilanti, was fired upon and one of the horses was killed. In spite of all these precautions the disease spread beyond Detroit; eleven died at Marshall. The disease took its victims quickly; persons in excellent health were suddenly stricken with a feeling of uneasiness, shortly were consumed with burning fever and a craving for cold drinks, after which came vomiting, intestinal spasms, general debility, and death. The disease runs its course in three to five days. The modern treatment is to support the patient over this period by intravenously injecting fluids and minerals into the bloodstream and applying medications to keep the patient comfortable. But the treatment of the disease by the physicians of 1832 was to use counterirritants and astringents, which only made the patient worse. Among those who died at Detroit in 1832 was Father Gabriel Richard; he did not succumb to cholera but to exhaustion caused by nursing and comforting the sick. A second epidemic of cholera broke out in Detroit in 1834; it was of relatively short duration but more deaths occurred than in 1832. Seven percent of the city's population died in August. (178–79)

In another account of this 1832 epidemic, Conot (1974) describes how cholera spread from immigrant ships to Detroit:

Cholera thrived. It spread from sailing ship to seaboard city, from seaboard city to steamer, and from steamer to inland city. On July 4, 1832, the *Henry Clay* docked in Detroit with a contingent of 370 soldiers for the Black Hawk War. Several were already ill. The next day, the first died. The vessel was ordered to cast off. But as men were felled by the score, it made land near Port Huron, fifty miles above Detroit. About one hundred fifty of the troops deserted, and made their way back to Detroit. The disease had no difficulty taking root among the pioneers who, camping near the waterfront, drew their water and passed their water indiscriminately along the riverbank. On July 6, Detroit recorded its first death. . . . With autumn and colder weather the epidemic subsided. The summer of 1834, however, brought its recurrence. As the governor himself died, those people who could afford to leave dashed helter-skelter out of the city. (30)

Concern for waterborne disease epidemics continued and in 1873 the city of Detroit purchased land between Jefferson Avenue and the Detroit River, then well outside the city, to construct a pumping station and standpipe tower to meet the growing population's demand for drinking water. The thinking was that cleaner and safer water could be found outside the city limit.

In 1879, Detroit's Water Works were built on this fifty-six-acre waterfront property. It instantly became a Detroit landmark because of the unique eighty-five-foot brick standpipe tower and pumping station and because it served as a city park (figure 5.1). Residents called it Water Works Park (Marzejka 2000). It was a huge tourist attraction—a place you had to bring your out-of-town guests.

The standpipe tower was constructed to equalize water pressure throughout Detroit's water mains. The pumping station, at its peak, pumped 290 million gallons of water per day. However, for many residents of metropolitan Detroit its main attraction was the 202–step stairway to reach a balcony at the top, which offered spectacular views of the Detroit River, the park, and downtown Detroit. New pumping stations would eventually come on-line to equalize and maintain water pressure, but the standpipe tower would remain a tourist attraction until 1945 when the tower was razed because of safety concerns.

During the early 1900s the International Joint Commission (1918), a U.S.–Canadian commission established to resolve boundary waters problems, reported the occurrence of "winter cholera" and compiled survey data of bacterial contamination of the Detroit River. The commission reported that bacterial densities changed markedly from the head of the Detroit River

Figure 5.1. A late 1800s portrait of the eighty-five-foot standpipe tower and pumping station at Detroit's Water Works Park (Burton Historical Collection, Detroit Public Library).

(less than 5 colonies/100 ml) to its mouth (11,592 colonies/100 ml) and that bacterial contamination was most pronounced close to shore.

Typhoid fever also became a major concern in the late 1800s and early 1900s, especially during the winter and spring months. Typhoid fever is an acute infectious disease caused by consumption of water or food contaminated with the bacterium *Salmonella typhi*. Symptoms start with a fever, which goes up over a three-day period, followed by severe headache and abdominal pain. If treatment is not obtained, approximately one in five people can die (Nester et al. 1973).

As far back as the 1880s there was public concern for protection of drinking water supplies. The American Public Health Association Committee on Pollution of Water Supplies (1888) concluded the following:

Many of our public water supplies contain sewage, and its harmfulness in a general way is unquestioned even by those who have a financial interest in them. Yet there appears to be a hesitancy to acknowledge the real, the specific, danger. Typhoid fever is present in all our cities, giving annual death rates of from 15 to 100 and over in every 100,000 of the population; but in the enumeration of its causes its prevalence is ascribed to many insanitary conditions before mention is made of the public water supply. . . . [But] we cannot shut our eyes to the relation that exists between sewage in our streams and typhoid fever in the cities that are supplied by them.

Bacterial contamination of Detroit's water supply, which came from the Detroit River, was identified as the cause of a typhoid fever epidemic in 1913, when 153 people died from this disease (International Joint Commission 1951). As a result, Detroit began disinfecting its water supply with chlorine in 1916.

During this time there was no sewage treatment. In a major international study of sewage pollution and public water supplies, the International Joint Commission (1949) concluded: "A potential menace is present where waters polluted to the extent of these are used for domestic purposes. They are in such condition that they cannot be safely used as a potable supply without complete and continuously effective treatment."

There were also other reports of waterborne disease epidemics. For example, an epidemic of gastroenteritis occurred in Detroit in 1926 when 45,000 people were affected and a total of 1,300,000 people exposed (International Joint Commission 1951). This incident was reported to also have been traced to sewage pollution of the water supply despite the filtration and chlorination.

To help ensure the safety of its public water supply, Detroit moved its municipal water intake from along the mainland shore to the opposite side of the U.S. channel along Belle Isle in 1932. It would not be until 1940 that Detroit would establish a primary waste water treatment plant (i.e., treatment made up of screening, settling, and disinfection). As public water treatment processes improved and Detroit moved its intake across the channel to Belle Isle, typhoid fever mortality was eventually eliminated (figure 5.2). However, bacterial contamination of the Detroit River was still occurring. Van Coevering (1948) told the following story of how people were still at risk:

Choosing the unusual to attract the attention of visitors to the Navy Club, a Detroit girl went swimming in the Detroit River at the foot of Griswold in June

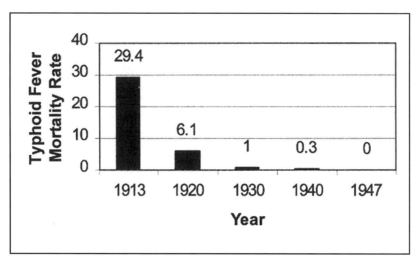

Figure 5.2. Typhoid fever mortality rates (per 100,000) in Detroit, 1913–47 (International Joint Commission 1951).

> 1943. A Board of Health inspector, on a routine inspection along the water front, spotted her. He called out to her to come out of the water. When she disregarded him, he went to find a policeman. She swam away. . . . How long she was in the water is not known. It is recorded, however, that she was taken from her home to Grace Hospital three weeks later. Her case was diagnosed as typhoid fever. (10)

The International Joint Commission (1951) voiced similar concerns and the following word of caution:

> The fact that typhoid fever from public water supplies has virtually disappeared must not be construed to mean that pollution has diminished. On the contrary, pollution of these waters has been steadily increasing, but water purification processes have made notable advances and, in general, have been able to cope with the problem of producing a safe water, even under adverse circumstances. (58)

In an effort to continuously ensure high-quality drinking water, Detroit's Board of Water Commissioners started looking in 1966 for a new source of water remote to the location of the Detroit River intakes. The board decided on Lake Huron. The Lake Huron Water Treatment Plant draws its water from a pipe six miles out in southern Lake Huron and pumps water to residents throughout southeast Michigan. The plant started pumping water in

1974, expanding the Water Department's total capacity to over 1,500 million gallons per day.

The Next Challenge of Controlling Combined Sewer Overflows

Despite the fact that typhoid fever epidemics have been eliminated, there continues to be concern over bacterial and viral contamination from combined sewer overflows (CSOs) to the Detroit River (i.e., where storm and sanitary sewers are combined and result in the discharge of raw sewage during rainfall events). For example, the U.S.–Canadian Remedial Action Plan for the Detroit River (Michigan Department of Natural Resources and Ontario Ministry of Environment 1991) has concluded the following:

> The river is significantly influenced by storm water runoff and combined sewer overflows. The sewer systems in Detroit and Windsor convey combined sanitary, industrial, and storm waste water. Excess combined waste water is discharged through combined sewer overflows directly to the Detroit River and its tributaries during rain events to protect the treatment plants from excessive hydraulic loadings. There are approximately 77 CSOs which discharge directly to the river (56 from Michigan, 21 from Ontario) and an additional 168 CSOs that discharge to the Rouge River and 11 CSOs which discharge to the Ecorse River (a Michigan tributary to the Detroit River). The total discharge from the City of Detroit's CSOs to the Detroit River was estimated to be an average of 34 million gallons per day in 1979. (4)

The Detroit Water and Sewerage Department treats the waste water of three million people from 78 communities and operates and maintains 3,500 miles of sewer lines, which carry rain water and waste water. If all the sewer lines were placed end to end in a straight line, the length would be greater than the distance from Boston to Los Angeles.

To address the CSO problem affecting the Detroit River, Detroit has adopted a Long-Term CSO Control Plan to reduce overflows of raw sewage. The four major elements of the plan include:

- controlling the amount of rain water that can get into the sewer system
- storing waste water in existing sewer pipes during storms (i.e., in-system storage)
- expanding the existing waste water treatment plant so that it can treat more of the combined sewage flow

- constructing large facilities to store and treat the combined sewage (i.e., end-of-pipe treatment)

The rain water control program and the waste water treatment plant expansion began in 2000. In-system storage and end of pipe treatment began in 2001. It is expected that it will take up to twelve years to fully implement all these measures at a cost of over $1 billion. Other downriver communities and ones up in the watersheds are undertaking similar programs to address CSO and urban runoff problems.

The Future

Today, Detroit's Water and Sewerage Department provides drinking water to over 4 million residents (43 percent of Michigan's population) in 125 communities. Detroit's water supply system includes three intake facilities (i.e., Belle Isle, southern Lake Huron, and near Fighting Island) and five water purification plants. It is well recognized that Detroit provides one of the highest quality, most reliable, economical water supplies in North America. Indeed, citizens of southeast Michigan are blessed with this water supply. In addition, the Detroit water system is going through a ten-year, $4 billion, capital improvement program that began in 2000.

However, drinking water supplies are still vulnerable to contamination. For example, contamination of Milwaukee's water supply with *Cryptosporidium* (i.e., a waterborne parasite that causes flu-like symptoms when ingested) occurred in 1993. Approximately 370,000 residents suffered from the largest outbreak of cryptosporidiosis ever recorded and 47 people died. Such experiences demonstrate that every effort must be expended to provide continuously effective water treatment in order to sustain the quality of our drinking water and prevent similar problems from occurring in southeast Michigan.

LITERATURE CITED

American Public Health Association Committee on Pollution of Water Supplies. 1888. Report presented at the November 20–23 Annual Meeting, Milwaukee.
Conot, R. 1974. *American Odyssey.* New York: William Morrow.
Farmer, S. 1890. *History of Detroit and Wayne County and Early Michigan.* Detroit: Silas Farmer and Company.
International Joint Commission. 1918. "Final Report of the International Joint Commission on the Pollution of Boundary Waters Reference." Washington, DC: U.S. GPO.

———. 1949. "Pollution of International Boundary Waters: 1946–1948" Investigation. Washington, DC and Ottawa.

———. 1951. "Pollution of Boundary Waters." Washington, DC and Ottawa: IJC.

Marzejka, L. J. 2000. Detroit's Water Works Park: A Gateway to the Past. *http://www.detroitnews.com.*

May, G. S. 1995. *Michigan: A History of the Wolverine State.* Grand Rapids, MI: Eerdmans.

Michigan Department of Natural Resources and Ontario Ministry of Environment. 1991. Stage 1 Remedial Action Plan for the Detroit River. Lansing, MI and Sarnia, Ont.

Nester, E. W., C. E. Roberts, B. J. McCarthy, and N. N. Pearsall. 1973. *Microbiology: Molecules, Microbes, and Man.* New York: Holt, Rinehart, and Winston.

Nolan, J. 2000. *How the Detroit River Shaped Lives and History. http://www.detroitnews.com.*

Van Coevering, J. 1948. *Save Michigan's Streams.* Detroit Free Press.

⤇ 6 ⤆

THE PUBLIC OUTCRY OVER OIL POLLUTION OF THE DETROIT RIVER

John H. Hartig and Terry Stafford

By the end of the 19th Century Detroit was a major manufacturing center. Detroit's manufacturing machine produced railroad cars, stoves, pharmaceuticals, furniture, cigars, and shoes. There were two primary reasons why Detroit became a major manufacturing center. First, Detroit had a prosperous ship building industry and the Detroit River provided the perfect transportation route to the north, west, and east. Secondly, raw materials like wood, iron ore, and brine were in plentiful supply and in relatively close proximity.

Detroit's first car company started in 1899. However, it did not take long for automobile manufacturing to take off. Henry Ford believed that cars should be affordable to everyone. To help achieve that goal, he created more efficient manufacturing systems, including assembly lines. By 1913 the industry grew to the point where there were 43 different automobile companies operating in the Detroit area. In 1914, Henry Ford announced that pay for an eight-hour shift in his Highland Park plant would be $8 per day (Cowles 1975). This drew a huge crowd of over 10,000 that had to be dispersed with fire hoses. The "Industrial Revolution" was well underway.

When World War I came Detroit focused on building enormous amounts of war machinery. No one seemed to care about the pollution associated with the war machine. As noted in chapter 5, there was concern for bacterial contamination of water supplies because of waterborne disease epidemics, but no real outrage over industrial pollution. During World War II Detroit became known as the "free world's arsenal" and the nation again

Figure 6.1. An oil-soaked canvasback duck taken from the Detroit River in March 1960, compared to a normal one with natural coloration (State Archives of Michigan).

centered its people, energies, and resources on arms production. Nobody had time to worry about the Detroit River and that mindset continued for several years after the war ended.

Detroit's apathy for industrial pollution of the Detroit River ended in 1948 when over 11,000 ducks died due to oil pollution. The winter of 1948 was a very cold one, allowing the river to freeze over. That left only a few areas of open water. Ducks over-wintering on the Detroit River headed for these openings that were filled with oil. The first massive duck kill due to oil pollution occurred and people were outraged (figure 6.1).

The Effects of Oil Pollution

The Great Lakes are an important waterfowl migration corridor because they are situated at the intersection of the Atlantic and Mississippi Flyways. It is estimated that three million ducks, geese, swans, and coots migrate annually through the Great Lakes region (Manny et al. 1988).

The Detroit River and western Lake Erie are extensive feeding and nesting areas for waterfowl (table 6.1). Wildlife biologists have recorded 27 species of waterfowl that frequent the coastal wetlands, including numerous species of ducks, coots, geese, and swans. Indeed, the national and international importance of this area is manifested in the 1961 Congressional designation of the 463–acre Wyandotte National Wildlife Refuge in the lower Detroit River. The Canard River Marsh Complex on the Ontario side has received a comparable designation in Canada. These habitats are unique because they are some of the main resting and feeding areas for canvasbacks migrating from their nesting grounds in Manitoba, Canada to the east coast (Metropolitan Affairs Coalition 2001). However, many over-winter in the area as well.

The lower half of the Detroit River is more intensively used by waterfowl than the upper half because the channels are divided and between these channels are shoals with dense stands of aquatic plants and an abundance of food for waterfowl. Just ask any duck hunter or naturalist about how important and unique the lower Detroit River is to waterfowl.

Prior to the mid-1930s a waterfowl mortality problem did not exist because the shallow, food-rich, lower half of the Detroit River froze from bank to bank and the narrow, deep, ice-free, upper half supported little food as an incentive for ducks to remain during the winter (Miller and Whitlock 1948). Then more and more industries, attracted by an almost unlimited supply of water, developed along the western shore of the lower river. Effluent from these industries warmed the water to the extent that parts of the

TABLE 6.1
Estimated Waterfowl Use of the Detroit River and Western Lake Erie Between the Late 1940s and Early 1960s

	Waterfowl Use	
Time Period	Average number of waterfowl	Range of number of waterfowl
Winter	48,000	6,000–116,000
Spring	55,000	28,700–140,000
Fall (pre-hunting season)	23,000	14,000– 38,300
Fall (mid-hunting season)	123,000	23,900–325,100
Fall (post-hunting season)	117,000	15,200–188,100

Source: U.S. Department of Health, Education, and Welfare 1962.

lower Detroit River were always ice-free (Miller and Whitlock 1948). Migrant waterfowl remained because of the open water and food. In many years, upward of 50,000 waterfowl remained each winter.

During the 1940s, oil was being released into the Detroit River in vast quantities. This oil pollution eventually became a major cause of mortality of waterfowl, which over-wintered and rested during migrations on the lower Detroit River (Hunt 1961). Waterfowl mortality frequently occurred during periods when open water was restricted by heavy ice cover (table 6.2). The largest of these winter duck kills occurred in 1948, 1960, and 1967 when 11,000 ducks, 12,000 ducks, and 5,400 ducks died, respectively. Ducks over-wintering on the Detroit River head for open water. This open water was the same place that during that time contained oil (figure 6.2).

Effects of oil on waterfowl are well known:

- oil mats feathers or fur permitting exposure of cold water to skin
- starvation may result due to reduced mobility, either swimming or flying
- actual drowning may occur
- sickness may result due to ingestion of oil (in general, small amounts of oil can be fatal to ducks)

Miller and Whitlock (1948) provide the following account of the 1948 winter duck mortality in the lower Detroit River:

TABLE 6.2
Waterfowl Mortality in the Detroit River Due Primarily to Oil Pollution

Year	Estimated Waterfowl Mortality
1948	11,000
1949	76
1950	871
1951	250
1952	1,000
1953	345
1954	238
1955	2,600
1956	191
1960	12,000
1967	5,400

Source: Hartig and Stifler 1979. Compiled from information from the U.S. Department of Health, Education, and Welfare 1962.

Figure 6.2. Oil pollution in the lower Detroit River during the winter of 1960 (Michigan Department of Natural Resources).

Shortly after the middle of January, the winter's worst period of near- and sub-zero temperatures had driven the ducks from their usual feeding grounds into the small channels and other small ice-free areas along the west bank of the river. It is in this area that industrial pollution is heaviest. Miles of busy factories representing one of the world's most concentrated industrial areas, making automobiles, steel, chemicals, etc., line the west bank of the river and its tributaries. Industrial wastes find their way into the river. One of the worst of these wastes is oil of various types. Oil adheres to the feathers of the ducks and gradually saturates them so that the insulating effect of the feather coat is broken down and the birds are subjected to the chilling effect of the cold weather. In addition, the extra weight and interference with the feather pattern of the wings greatly hamper the power of flight. Oil-soaked birds move about much less than normal birds and frequently leave the water and their normal feeding grounds. They seem to spend much of their time and energy in removing the oil from their feathers by constant preening. Toward the end of January a serious situation had developed. It appeared to Game Division investigators that a combination of factors was responsible for the alarming mortality rates, principally among canvasback ducks. In mid-January a heavy oil flow had come down the west bank of the river and several thousand ducks had been affected by it in varying degrees. . . . Finally, about mid-February, another wave of oil came down the river from some unknown source. This oil flow was sufficient to contaminate nearly all of the feeding areas still free of ice. For a period of three or four days the birds were forced to disperse to other parts of the river or leave the water completely. Many of them braved the oil and a good percentage of those that did failed to survive the combination of oil, intense cold, and lack of food. (11, 15)

Control of Oil Pollution

Michigan Department of Natural Resources staff attributes the 1948 winter duck kill due to oil pollution as the catalyst for the industrial pollution control program in Michigan. Truck loads of oil soaked carcasses of dead ducks were driven to Lansing by angry downriver sportsmen and dumped on the Capitol lawn in 1948 (Cowles 1975). In addition, conservationists and duck hunters delivered ducks that had died due to oil pollution in the Detroit River to the Michigan Capitol in Lansing to further heighten awareness of the oil pollution (figure 6.3).

Government leaders could no longer ignore the evidence and were forced by the public to institute industrial pollution control programs targeted first at oil pollution, but later at many more pollutants.

During 1949 the Michigan Legislature revised the water pollution control statute (Cowles 1975). That new amendment established the Michigan Water Resources Commission, expanded the definition of pollution, and required state approval of all new uses of state waters. The 1948 duck kill due to oil pollution in the Detroit River was the proverbial "last straw that broke the back of the overburdened camel" in the fable.

The necessary changes to control oil pollution started slowly. During the winter and spring periods from 1950 to 1955, U.S. Department of Health, Education, and Welfare (1962) recorded that oil slicks were present on the Detroit River one-third of the time. Early efforts focused primarily on industries as the major source of oil pollution in the Detroit River (International Joint Commission 1951; table 6.3).

Figure 6.3. Oil-soaked carcasses of waterfowl from the lower Detroit River that were delivered to the Michigan Capitol in protest of oil pollution in the Detroit River, 1948 (State Archives of Michigan).

By the 1960s it was well recognized that a variety of sources contributed to the oil pollution problems of the Detroit River, including industries, municipal wastewater treatment plants, government installations, combined sewer overflows, and shipping (International Joint Commission 1968). Records from June 30, 1960 through January 5, 1962 indicated that at least 36 observations of oil on the Detroit River in the vicinity of or discharging from sewers (U.S. Department of Health, Education, and Welfare 1962). Fifteen of these oil discharges were from steel plants, five from municipal sewers or drains, five from automobile plants, five from coal distillation processes, two from brass production, and four from miscellaneous sources. Although small amounts of oil were being discharged in many cases during that time, the dispersion of oil into a slick could cover large areas of the Detroit River. Current and wind conditions tended to concentrate oil slicks in marinas and harbors where damage to pleasure boats occurred. In addition

TABLE 6.3
Known Industrial Sources of Oil Pollution to the Detroit River During the Early 1950s (International Joint Commission 1951)

- U.S. Rubber Company
- Michigan Consolidated Gas Company
- American Brass Company
- Revere Copper and Brass Company
- Semet-Solvay Division of Allied Chemical & Dye Corporation
- Barrett Division of Allied Chemical & Dye Corporation
- Great Lakes Steel Corporation
- Ford Motor Company
- Darling and Company
- Fuel Oil Corporation
- Murray Corporation of America
- Wyandotte Chemicals Corporation
- Warner G. Smith Company
- Firestone Rubber and Metal Products Company
- Sharples Chemicals Incorporated
- Socony-Vacuum Oil Company
- Ford Motor Company of Canada

Source: International Joint Commission 1951.

to damage to pleasure boats and degradation of aesthetic conditions at bathing beaches, certain oils contributed to fire hazards and taste and odor problems in water supplies (International Joint Commission 1968).

Again, it was the effects of oil pollution manifested in winter duck kills that heightened awareness and elevated the profile of the need to control oil discharges. Reductions in oil discharges became equally as dramatic as the frequency and extent of winter duck kills due to oil pollution. For example, between the late 1940s and early 1960s there was a 97.5 % reduction in oil discharges to the Detroit River (figure 6.4). A follow-up study performed by the Michigan Department of Natural Resources (1977) reported that there had been an additional 80% decrease in point source discharges of oil between 1963 and 1976. However, that study reported that storm and sanitary sewers still contributed significant amounts of oil during heavy rainfall events. Today, winter duck kills due to oil pollution on the Detroit River have been eliminated. The Detroit River has reclaimed its place as an ecological nesting a resting place for waterfowl with no adverse impacts from oil pollution.

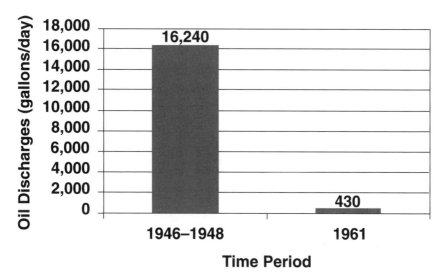

Figure 6.4. Reductions in oil discharges into the Detroit River between 1946–1948 and 1961 (U.S. Department of Health, Education, and Welfare 1962).

Concluding Remarks

The importance of the Detroit River as waterfowl habitat was celebrated in 2001 with the designation of the Detroit River as the first International Wildlife Refuge in North America. This will provide heightened awareness of the uniqueness of the Detroit River and the need to protect it from oil and other pollution, and from further habitat loss.

In recent years, oil in the Detroit River has been negligible due to the fact that all industries and municipalities have discharge limits for oil and grease. Only occasional losses have generally occurred because of state and federal enforcement programs. Small oil spills (less than 100 gallons) have occurred, but have been cleaned up by the U.S. Coast Guard and other governmental and industrial partners.

One exception to this occurred in April 2002 when a large oil spill occurred in the Rouge River that eventually flowed into the Detroit River. The U.S. Coast Guard and partners recovered over 40,000 gallons of oil. This was the largest oil spill to occur in the Great Lakes and their connecting channels in the last 12 years. This oil spill affected the entire U.S. shoreline of the Detroit River and a small portion of the Lake Erie shoreline. Cleanup costs exceeded $3 million. The U.S. Fish and Wildlife Service reported that

10 ducks and geese, and two turtles, died as a result of the oil pollution. Commercial navigation had to be closed on the Rouge River for almost two weeks. The magnitude of this oil spill reminds us of the need for continuous and vigorous oversight of industrial and municipal discharges.

Every effort must be placed to prevent such spills in the future and to ensure that there is an early warning system in place to protect our Detroit River. The old adage that "an ounce of prevention is worth a pound of cure" really holds true on this issue. The good news is that massive, winter duck kills due to oil pollution have been eliminated. The bad news is that the April 2002 Detroit River oil spill was substantial and resulted in the loss of 10 birds and two turtles. We need to treat the Detroit River like our home and not defile it with oil and other contaminants!

LITERATURE CITED

Cowles, G. 1975. Return of the River. Michigan Natural Resources Magazine, January-February issue, Lansing, MI

Hartig, J. H. and M. E. Stifler. 1979. Water Quality and Pollution Control in Michigan. Michigan Department of Natural Resources Publication Number 4833-9803, Lansing, MI

Hunt, G. S. 1961. Waterfowl losses on the lower Detroit River due to oil pollution. Publication No. 7, Great Lakes Research Division, University of Michigan, Ann Arbor, MI

International Joint Commission. 1951. Pollution of Boundary Waters. Washington, D.C. and Ottawa.

International Joint Commission. 1968. Pollution of the Detroit, St. Clair, and St. Marys Rivers. Washington, D.C. and Ottawa.

Manny, B. A., T. A. Edsall, and E. Jaworski. 1988. The Detroit River, Michigan: An Ecological Profile. U.S. Fish and Wildlife Service, Biological Report 85 (7.17), Ann Arbor, MI

Metropolitan Affairs Coalition. 2001. A Conservation Vision for the Lower Detroit River Ecosystem. Detroit, MI

Michigan Department of Natural Resources. 1977. The Detroit River: 1966–1976: A Progress Report. Publication Number 4833-9438, Lansing, MI

Miller, H. J. and S. C. Whitlock. 1948. Detroit River ducks suffer heavy losses. Michigan Conservation 17 (4): 11, 15.

U.S. Department of Health, Education, and Welfare. 1962. Pollution of the navigable water of the Detroit River, Lake Erie and Their Tributaries within the State of Michigan. Detroit, MI

DETROIT'S ROLE IN REVERSING CULTURAL EUTROPHICATION OF LAKE ERIE

Jennifer Panek, David M. Dolan, and John H. Hartig

The first ecological changes in Lake Erie occurred around 1850 when forestland was cleared for agricultural purposes (Harris and Vollenweider 1982). Since then at least parts of Lake Erie have exhibited what scientists call mesotrophic status or moderate nutrient enrichment and production of algae. The main evolution toward what scientists call eutrophic status or high nutrient enrichment and overproduction of algae occurred during the twentieth century. By the early 1960s Lake Erie was exhibiting accelerated or cultural eutrophication (Beeton 1961). Eutrophication is a natural aging process of lakes, but it can be accelerated by human activities. Cultural eutrophication refers to the accelerated aging of a lake caused by elevated nutrient loadings from human activities.

During the 1960s the public at large recognized that Lake Erie was highly eutrophic. Lake Erie had changed such that:

- algal blooms covered large areas of the lake during the summer months
- attached algae called *Cladophora* covered most rocky and man-made structures (figure 7.1)
- decomposing algae, which had washed up on bathing beaches, had to be removed by bulldozers
- blue-green algae were causing taste and odor problems in some municipal water supplies
- dissolved oxygen was absent from many deepwater areas of the lake

Figure 7.1. Attached algae called *Cladophora* along the shoreline of South Bass Island in western Lake Erie during the 1960s (C. E. Herdendorf, Ohio State University).

Major changes occurred in both plant and animal life. Oxygen depletion in the lower waters of Lake Erie caused major changes in the fishery. Pollution-tolerant fish like carp and goldfish became more abundant. Mayfly larvae, which live in sediments, could not survive because of the lack of oxygen. In fact, Vaughan and Harlow (1965) report that by the mid-1960s mayflies were completely eliminated from Lake Erie. Pollution-tolerant sludge worms took the place of mayflies.

The media picked up on the news of eutrophication of Lake Erie and erroneously declared it "dead." One of the symptoms of the eutrophication of Lake Erie was that, as noted above, oxygen was depleted from the deeper waters of the lake and this caused some fish to die. These fish would windrow ashore and wash up with the decomposing algae. The public perceived the lake as "dead," but the problem was that it was highly productive as a result of elevated phosphorus inputs.

Scientists refer to phosphorus as the limiting nutrient of freshwater lakes. That means that phosphorus is the nutrient in least supply relative to the needs of algae. In general, the more phosphorus that is added to Lake Erie, the greater the production of algae. The greater the production of algae, the greater the water quality problems.

During the mid-1960s, total phosphorus loading to Lake Erie was approximately 62,000 kg/day (U.S. Federal Water Pollution Control Administration 1968). Seventy-two percent of the loading was from municipal wastes, 17 percent from rural runoff, 4 percent from industrial wastes, and 7 percent from urban runoff. Of the phosphorus coming from municipal wastes, 66 percent came from phosphorus-based detergents.

As a result of widespread concern for the eutrophication of Lake Erie and the research identifying phosphorus as the limiting nutrient, the United States and Canada established a comprehensive phosphorus control program to restore and protect all of the Great Lakes. This binational program became the cornerstone of the Great Lakes Water Quality Agreement that was signed on April 15, 1972, by President Richard Nixon and Canadian Prime Minister Pierre Trudeau (United States and Canada 1972). In this agreement the U.S. and Canadian governments set phosphorus loading targets. All municipal wastewater treatment plants that handled more than or equal to one million gallons per day (3,800 cubic meters per day) were required to reduce effluent total phosphorus concentrations to less than or equal to 1 mg/L. In addition, all jurisdictions were encouraged to adopt phosphorus detergent bans (i.e., limiting the phosphorus content of laundry detergents to less than or equal to 0.5 percent by weight). It was determined that these two efforts alone would be enough to maintain the trophic status of Lakes Superior, Michigan, and Huron. However, in order to alleviate taste and odor problems in municipal water supplies from Saginaw Bay, eliminate nuisance algal conditions in Lakes Erie and Ontario, and restore the year-round oxygenated conditions in the hypolimnetic waters (i.e., deep waters) of the central basin of Lake Erie, additional reductions in phosphorus loading would be required. Phosphorus management plans and implementation schedules were required to meet phosphorus-loading reduction targets. In these cases, nonpoint source (i.e., diffuse sources of pollution, such as runoff from precipitation, that do not enter a river at a fixed point) controls would be needed to reduce agricultural and urban runoff of phosphorus.

The Detroit Wastewater Treatment Plant

The single largest contributor of phosphorus to Lake Erie was the Detroit Wastewater Treatment Plant (WWTP)(figure 7.2). The initial phase of the Detroit WWTP was completed in 1940. The plant was constructed as a regional system and designed to meet the needs of an increasing population throughout the metropolitan area as well as to provide primary treatment (no phosphorus removal) to the sanitary waste water of the city and suburban communities.

Figure 7.2. An aerial photograph of the Detroit Wastewater Treatment Plant, one of the largest in North America (Detroit Water and Sewerage Department).

Detroit's WWTP system has provided wholesale service to an increasing number of surrounding municipalities since the initial construction of the plant in 1940. The greatest growth period for wholesale municipal customers occurred during the late 1950s and early 1960s. However, it would not be until the late 1960s when the cultural eutrophication of Lake Erie became widely known that Detroit and other areas would come to recognize their role in the deterioration of Lake Erie and the need for more advanced treatment and phosphorus control. Major WWTPs are defined as plants, which treat one million or more gallons of municipal waste per day. The Detroit WWTP grew such that it handled over 700 million gallons per day and was the largest plant in the Great Lakes Basin (table 7.1). As a result, the Detroit WWTP was the single largest source of phosphorus to Lake Erie. In 1980 the Detroit WWTP accounted for 40–45 percent of the municipal phosphorus loading to Lake Erie (Hartig 1983).

Detroit's efforts to control municipal phosphorus inputs are representative of similar efforts around the Great Lakes (Dolan 1993). In 1970 polymer and pickle liquor feeding facilities were added to precipitate out phosphorus (Hartig 1983). Pickle liquor (ferrous chloride) was obtained from local steel mills and pumped or fed by gravity into interceptor sewers, while polymer was injected into channels leading to the primary clarifiers. Aeration facilities for secondary treatment were constructed during 1973–76. Through this process, ferrous chloride is converted to ferric chloride, which has been found to more effectively precipitate phosphorus. During 1979–80 staff of the Detroit WWTP implemented an alternative sludge removal process, which increased sludge handling capability and indirectly increased the plant's ability to remove phosphorus.

TABLE 7.1
Detroit Wastewater Treatment Plant and Service Area Statistics

Service area: 821 square miles in Wayne, Oakland, and Macomb Counties
Number of communities served: 76
Population served: 3 million people (31 percent of state population)
Number of miles of trunk and lateral sewers: 2,900 miles
Number of pump stations: 13
Interceptor sewers: 3 in the city of Detroit and 1 outside
Flow: approximately 700 million gallons per day
Level of treatment: activated sludge secondary treatment with phosphorus removal

Source: Detroit Water and Sewerage Department, www.dwsd.org/index.htm

It should also be noted that two statewide phosphorus control initiatives were implemented (Hartig 1983). In 1971, Michigan enacted a phosphorus limitation of 8.7 percent by weight on all cleaning agents. Michigan's phosphorus detergent ban was implemented in 1977, restricting the phosphorus content of household laundry detergents to no greater than 0.5 percent by weight.

The combined influence of these phosphorus control efforts can be seen in figures 7.3a and 7.3b. The result was a more than 90 percent reduction in phosphorus concentration and loading from the Detroit WWTP. Similar reductions occurred in other WWTPs; however, due to Detroit WWTP's 700 million gallon per day flow, this effort had the most significant impact on Lake Erie. The Detroit WWTP was the single largest contributor to the reversal of cultural eutrophication of Lake Erie during the 1970s and 1980s.

It should also be noted that as the Detroit WWTP expanded its municipal customers and service area there were problems in consistently and adequately operating the plant. These problems climaxed in a 1977 federal consent judgment, which outlined the specific deficiencies, areas requiring improvement, and target dates for achieving compliance. A full-scale evaluation of the plant was performed in 1979 to help ensure adequate operation of the plant. It would not be until 1981, when the construction and modification of secondary settling tanks were completed, that the plant would achieve consistent operation sufficient for secondary treatment standards.

Impact on Lake Erie

Noticeable changes began to occur in Lake Erie by the mid-1980s. Because the Detroit River accounts for approximately 93 percent of the inflow to Lake Erie and because of the substantial reductions in phosphorus loadings from the Detroit and other WWTPs to the Detroit River, Lake Erie responded first with decreased phosphorus concentrations. For example, there was approximately 60–70 percent reduction in phosphorus concentration measured at the Kingsville Water Intake on the north shore of the western basin (figure 7.4). In general, phosphorus concentrations declined from a level indicative of a highly eutrophic state to a level more indicative of a mesotrophic state. However, it should be noted that concentrations continue to be quite variable because of the shallow nature of the basin, which allows wind to resuspend sediment and nutrients like phosphorus. Scientists refer to this process of wind-induced resuspension of sediment and phosphorus as internal loading.

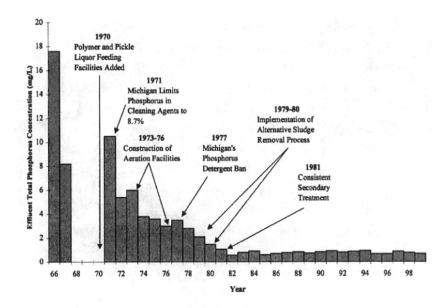

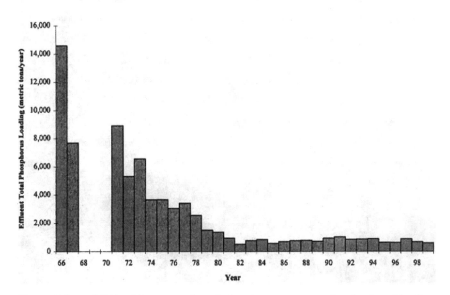

Figures 7.3a and 7.3b. Mean annual total phosphorus concentration (top) and loading (bottom) from the Detroit Wastewater Treatment Plant to the Detroit River, 1966–99.

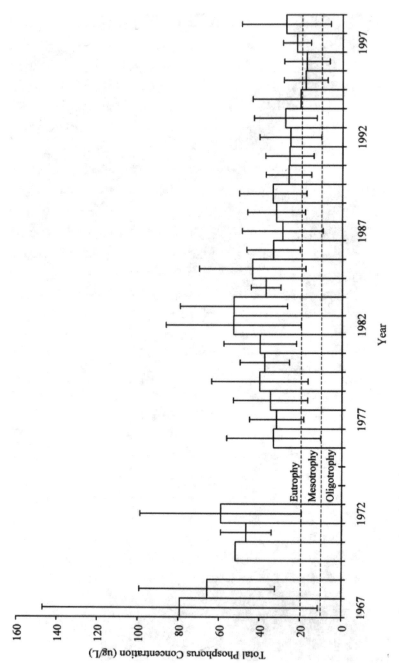

Figure 7.4. Mean annual total phosphorus concentrations (ug/L) in the western basin of Lake Erie as reported by the Union Water Intake at Kingsville, Ontario, 1967–98 (data collected by the Ontario Ministry of the Environment).

The first significant ecological change was an 89 percent decline in the blue-green alga *Aphanizomenon flos-aquae* between 1970 and 1983–85 (Makarawicz and Bertram 1991). These algae had historically caused taste and odor problems in some municipal water supplies. The relative proportion of blue-green algae decreased while the relative proportion of diatoms (i.e., a group of algae that is indicative of clean water) increased (Neilson et al. 1995). This shift in phytoplankton is indicative of improved water quality in Lake Erie in direct response to reduced phosphorus loadings.

During the 1950s and 1960s, when elevated phosphorus inputs led to high production of algae, the algae would eventually sink to the bottom of the lake and be decomposed by bacteria, using up dissolved oxygen in the process. By the 1980s, the lower phosphorus levels and algal productivity had set the stage for the improvements in oxygen levels of the lake. Oxygen depletion rates decreased in the deeper waters of Lake Erie and they became less variable. Dissolved oxygen was not being depleted from the deeper waters as quickly because algal productivity had decreased due to reduced phosphorus loadings. The central basin of Lake Erie continued experiencing periods of anoxia (i.e., without oxygen) during the summer months throughout the 1980s and 1990s (Neilson et al. 1995). However, oxygen returns to the central basin in the fall and remains plentiful until the following summer (Neilson et al. 1995).

Lack of oxygen was believed to have caused the extinction of mayfly larvae *(Hexagenia)* in Lake Erie by the mid-1960s. By 1980, dissolved oxygen levels had increased, resulting in the reappearance of mayfly larvae (Burns 1985). Their return started out slow but has recently been more dramatic. For example, Jentes (1998) reports that in one Lake Erie survey *Hexagenia* larvae increased from 34 per square meter in 1995 to 404 per square meter in 1997.

Improvements in zooplankton community structure and composition also occurred (Makarawicz and Bertram 1991). The fishery responded with a shift from pollution-tolerant species to ones that require cleaner water and higher oxygen levels (Colby, Lewis, and Eshenroder 1991).

By the late 1980s, the walleye population in Lake Erie began increasing. While in the previous decade their numbers were very low, Colby, Lewis and Eshenroder (1991) report that the species once again had healthy numbers. The resurgence of the western Lake Erie walleye population in the 1980s is believed to be due to a ban on the commercial harvest of walleye (imposed because of mercury contamination in 1970) and several strong year classes resulting from favorable spawning conditions. This apparently resulted in a decline in several plankton-feeding fish species, which allowed large zooplankton species (water fleas called *Daphnia*) to increase in abundance. These large zooplankton species are also filter feeders, which graze on phytoplankton. This led scientists to conclude that improved water quality in western Lake Erie was the result of both bottom-up controls (i.e., reduced phosphorus loadings) and top-down controls (i.e., strong walleye populations that created a cascading effect on phytoplankton densities) (Nicholls 1999).

Another important part of the story was the introduction of zebra mussels in Lake Erie in 1988. In the early 1990s quagga mussels appeared. Both of these mussels were probably introduced from ballast water discharged from transoceanic ships. The rapid spread of these mussels led to major changes in Lake Erie. Zebra and quagga mussels are filter feeders, removing nearly all the phytoplankton and some small forms of zooplankton from the water, and thereby reducing the food supply for planktivorous fish.

Water clarity increased because the zebra and quagga mussels were filtering out the algae (Leach 1993). However, by removing phytoplankton, these mussels removed an important food source for larger zooplankton, which in turn are food for larval and juvenile fish and other plankton-feeding forage fish. Major changes in the Lake Erie food web resulted.

Concluding Remarks

The story of reductions in phosphorus loadings and the subsequent reversal of cultural eutrophication of Lake Erie is one of the greatest success stories of water resource management (table 7.2). Indeed, the U.S.–Canada phosphorus control program is heralded as an international model.

Blooms of *Microcystis,* a toxic blue-green alga, also occurred in Lake Erie during 1995, 1996, and 1997. This raises new concerns. The exact reasons for these blooms are uncertain. Did increased discharge of nutrients occur? Did zebra mussels change the water quality and favor productivity of blue-green algae? Will this lead to more taste and odor problems in drinking water supplies? We need focused research to answer such questions.

Recently there have also been concerns about the decline in the abundance of some of Lake Erie's most important fish, such as yellow perch, rainbow smelt, and walleye. These declines are thought to be strongly related to the combined effects of reduced phosphorus loadings and the invasion of zebra and quagga mussels. Research must be undertaken to understand and unravel the relative and cumulative effects of phosphorus control and mussel

TABLE 7.2
**Summary of Changes in Lake Erie Resulting from
Reduced Phosphorus Loadings**

Indicator	Change
Phosphorus concentration	60–70% reduction measured at the Union Water Intake, Kingsville, Ontario
Density of algae	(1) 42% decline in nearshore phytoplankton biomass between 1967 and 1975 (2) 89% decline in the blue-green alga *Aphanizomenon flos-aquae* between 1970 and 1983–85
Hypolimnetic oxygen levels	Decreased oxygen depletion rates and less variability
Density of mayfly larvae	Reappearance of mayfly larvae, including an order of magnitude increase in mayfly larvae between 1995 and 1997

Sources: Figure 7.3; Ontario Ministry of Environment; Nicholls et al. 1977; Makarawicz and Bertram 1991; Neilson et al. 1995; Jentes 1998.

invasion on the productivity of fish in Lake Erie (Lake Erie Committee 1998). Some sport-fishing interest groups have called for an increase in phosphorus loadings from wastewater treatment plants to be able to increase the amount of algae in Lake Erie and hence fish production. However, such increases in phosphorus loadings could cause unintended effects, such as a return of nuisance algal conditions in nearshore areas like bathing beaches, taste and odor problems in water supplies, increased abundance of mussels, and increased levels of contaminants that would be reduced during phosphorus removal at wastewater treatment plants. Therefore, the Lake Erie Committee (1998) of the Great Lakes Fishery Commission has recommended a "hold-the-line on phosphorus levels" position until there is clear scientific evidence that this would be an appropriate strategy. Currently that level of scientific certainty is impossible to achieve because major cuts have occurred in monitoring and research for both the Detroit River and Lake Erie. The major, poorly understood changes in Lake Erie have taught us that management programs, research, and monitoring must be sustained and closely coupled in order to achieve our goals for the Detroit River and Lake Erie.

LITERATURE CITED

Beeton, A. M. 1961. "Environmental Changes in Lake Erie." *Transactions of the American Fisheries Society.* 90:153–59.

Burns, N. M. 1985. *Erie: The Lake That Survived.* Totowa, NJ: Rowman and Allanheld.

Colby, P. J., C. A. Lewis, and R. L. Eshenroder, eds. 1991. *Status of the Walleye in the Great Lakes: Case Studies Prepared for the 1989 Workshop.* Great Lakes Fishery Commission Special Publication 91-1, Ann Arbor, MI.

Dolan, D. M. 1993. "Point Source Loadings of Phosphorus to Lake Erie, 1986–1990." *Journal of Great Lakes Research* 19.2:212–23.

Harris, G. P., and R. A. Vollenweider. 1982. "Paleolimnological Evidence of Early Eutrophication in Lake Erie." *Canadian Journal of Fisheries and Aquatic Sciences* 39:618–26.

Hartig, J. H. 1983. "Phosphorus Control Measures Pay Off for the Great Lakes." *Water Engineering & Management* 130.13:32–35.

Jentes, J. E. 1998. "Mayflies and Lake Erie." *Twine Line* 20.3:1.

Lake Erie Committee. 1998. *Phosphorus Targets Achieved in Lake Erie: Fisheries Agencies Recommend Holding-the-Line on Phosphorus Levels.* Great Lakes Fishery Commission Press Release, Ann Arbor, MI.

Leach, J. H. 1993. "Impacts of Zebra Mussel (Dreissena) on Water Quality and Fish Spawning Reefs in Western Lake Erie." In *Zebra Mussels: Biology, Impact, and Control,* ed. T. F. Nelepa and D. W. Schloesser, 381–97. Ann Arbor, MI: Lewis Publishers.

Makarawicz, J. C., and P. Bertram. 1991. "Evidence for the Restoration of the Lake Erie Ecosystem." *BioScience* 41.4:216–23.

Neilson, M., S. L'Italien, V. Glumac, and D. Williams. 1995. "Nutrients: Trends and System Response." Working Paper presented at State of the Lakes Ecosystem Conference, Chicago. EPA 905–R-95–015.

Nicholls, K. 1999. "Evidence for a Trophic Cascade Effect on North-shore Western Lake Erie Phytoplankton Prior to the Zebra Mussel Invasion." *Journal of Great Lakes Research* 25.4:942–49.

Nicholls, K. H., D. W. Standen, G. J. Hopkins, and E. C. Carney. 1977. "Declines in the Nearshore Phytoplankton of Lake Erie's Western Basin since 1971." *Journal of Great Lakes Research* 3:72–78.

United States and Canada. 1972. *Great Lakes Water Quality Agreement.* Washington, DC and Ottawa: International Joint Commission.

U.S. Federal Water Pollution Control Administration. 1968. *Lake Erie Report: A Plan for Water Pollution Control.* Washington, DC: U.S. Dept. of Interior.

Vaughan, R. D., and G. L. Harlow. 1965. *Report on Pollution of the Detroit River, Michigan Waters of Lake Erie, and Their Tributaries: Summary, Conclusions, and Recommendations.* U.S. Department of Health, Education, and Welfare, Public Health Service, Division of Water Supply and Pollution Control. Washington, DC: U.S. GPO.

MERCURY AND PCB CONTAMINATION OF THE DETROIT RIVER

Jennifer Read, Doug Haffner, and Pat Murray

When Rachel Carson released *Silent Spring* in 1962, the Detroit River was polluted with oils and grease and high levels of disease-causing bacteria. As Carson predicted, the darker menace of toxic chemicals also lurked beneath the surface. *Silent Spring* forever changed how we think about synthetic chemicals in our environment. Carson argued that modern agricultural and forestry practices overused pesticides, particularly DDT. She pointed out that these chemicals persist in the environment, accumulate in the bodies of animals and are transferred to humans ingesting their residues, and endanger the health of following generations (Carson 1962; Lear 1997).

Unlike highly visible environmental contaminants, such as oils, grease, and detergent foam, the impact of persistent chemicals such as mercury and polychlorinated biphenyls (PCBs) is insidious and long-term. In her work, Carson made it clear to a large and previously unaware segment of society that actions assumed to have no impact can have long-term implications. The book forced people to reexamine how they interact with their environment and reconsider how chemicals are approved, regulated, and used.

Silent Spring initiated a period during which researchers and resource managers around the world began to search for and understand how persistent chemicals impact ecosystems. There are now over fifty chemicals known to be in the Detroit River (table 8.1), which represents less than 0.1 percent of the synthetic chemicals known to exist. Although resource managers have made significant inroads in stopping their release, historically discharged chemicals are trapped in Detroit River sediments and recirculated when the

TABLE 8.1
Persistent Toxic Chemicals Known to Be in the Detroit River

Metallic Elements

Arsenic
Copper
Zinc
Cadmium
Mercury
Lead
Silver
Chromium
Nickel
Cobalt
Manganese

Pesticides

Aldrin
Dichloro-Diphenyl-Trichloroethane (DDT) and metabolites DDE, DDD, TDE
Dieldrin
Endosulfan I
Endosulfan II
Heptachlor
Lindane: a-BHC, b-BHC, c-BHC
Mirex
a- and g-Chlordane

Combustion Products

Acenaphthene
Anthracene
Benzo(a)anthracene
Benzo(a)pyrene
Benzo(b)fluoranthene
Benzo(g,h,i)perylene
Chlorobenzenes
Chrysene
Coronene
Dioxin
Fluoranthene
Fluorene
Indeno(1,2,3-cd)pyrene

Naphthalene
Perylene
Phenanthrene
Polycyclic Aromatic Hydrocarbons (PAHs)
Pyrene

Other Synthetic Chemicals

Bis(2-ethylhexyl)phthalate
Butyl benzyl phthalate
Dibenzo(a,h)anthracene
Dichlorophenol
Diethylphthalate
Di-n-butylphthalate
Dioctylphthalate
Heptachlorbenzo(e)pyrene
Hexachlorobenzene
Octachlorostyrene (OCS)
Pentachlorobenzene (QCB)
Pentachlorophenol
Polychlorinated Biphenyls (PCB)
polychloroterphenyl (PCT)
Tetrachlorophenol
Trichlorophenol

---

sediments are disturbed physically or taken up into the food web through biological activity. Scientists have also discovered that airborne chemicals, including mercury and PCBs, travel from sources many thousands of miles away and are deposited in the Detroit River.

Mercury and PCBs were among the first toxic chemicals of concern to be reported in the Detroit River. Researchers discovered elevated levels of mercury in Lake St. Clair walleye in 1969. In 1970, while searching for DDT residue in Great Lakes fish, scientists found elevated levels of PCBs. These two contaminants are good examples of the problems associated with the unconsidered release of toxic chemicals into ecosystems. While mercury occurs naturally and PCBs are manufactured by humans, both are persistent and bioaccumulative.[1] Scientists discovered both distributed throughout the Detroit River in the early 1970s and despite the fact that all known discharges were stopped soon after, both remain a problem today in the river ecosystem (see table 8.1). Toxic chemicals, such as mercury and PCBs, affect the fish and wildlife populations of the river, endanger the lives of human

Figure 8.1. A healthy brown bullhead (Todd Leadley).

beings using those resources, and help perpetuate the river's reputation as one of the most contaminated in North America (figures 8.1 and 8.2).

How was the river contaminated with mercury and PCBs and what is its status today? What have the governments done to address the problem of persistent toxic chemicals? What more should they do? What can you do to improve the health of the river? Through an examination of these questions the story of toxic chemicals in the Detroit River will be revealed.

The Mercury Crisis

Mercury is the silver liquid often referred to as quicksilver. It is not a synthetic chemical but a naturally occurring element most often associated with the ore cinnabar. It is volatile, readily vaporizing into the atmosphere. Mercury is a heavy metal that causes damage to organisms at low concentrations and accumulates in the food web. In its elemental form, mercury is relatively safe and must be inhaled to endanger living things. When elemental mercury is metabolized by bacteria, it is changed into an organic form, called methyl mercury, that is readily absorbed by organisms and, therefore, toxic. Given the toxicity of methyl mercury, during the early part of the twentieth century the agricultural industry readily adopted it as a fungicide to protect vulnerable seed grains. As a result, much more mercury was introduced into the

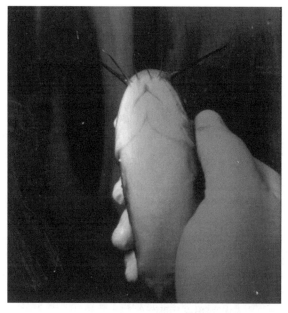

Figure 8.2. A bullhead with lip and skin tumors (Todd Leadley).

environment, causing an increase in the incidence of poisoning in fish, birds, and human beings (D'Itri 1972).

The most dramatic example of mercury poisoning occurred between 1953 and the late 1960s in Minamata, Japan, where effluent containing methyl mercury from a chemical factory accumulated in fish and caused 121 cases of poisoning of people who ate the fish; 46 died. In Sweden, the use of methyl mercury compounds as seed dressing caused a visible decline in the population of seed-eating birds, birds that preyed on seed-eating rodents, and expecially fish-eating birds by the 1950s. In the early 1960s, Swedish researchers found levels of several parts per million (ppm) in the tissue of freshwater fish.[2] The Swedish National Institute of Public Health determined that a level of 0.5 ppm in fish should be safe for human consumption because it was roughly 100 times less than the amount reported in the Japanese poisonings. Moving swiftly, Swedish authorities banned the use of methyl mercury as seed dressing and instituted fish consumption advisories (D'Itri 1972).

Alarmed by events in Europe and Asia, resource managers and health authorities in the Great Lakes Basin began to search for mercury in the Great Lakes ecosystem. Agriculture and industry in the basin employed processes similar to those European practices that had caused environmental contamination. Organic mercury dressing for grain seeds had been embraced by Canadian and U.S. farmers. Pulp and paper mills used mercury and methyl mercury as a slimicide, and it was a significant component of the chlorine manufacturing process.

In 1969 the Ontario Water Resources Commission began monitoring levels of mercury in St. Clair River sediment. The commission's concern stemmed from mercury known to be released by Dow Chemical of Canada from its chlor-alkali plant in Sarnia. The facility used mercury cell plants to manufacture chlorine and caustic soda and in the process released some of the mercury into the environment. Authorities estimated that plant effluent ran as high as 50 ppm at times, amounting to a release of approximately 91 metric tonnes of mercury into the St. Clair River. This mercury, in turn, contaminated Lake St. Clair, the Detroit River, and Lakes Erie and Ontario.

At the same time, the Canadian Wildlife Service (CWS) began to look for mercury contamination in fish and fish-eating birds in the vicinity of known mercury sources. A University of Western Ontario chemist, under contract with CWS, found elevated levels of mercury in St. Clair River sunfish and Lake St. Clair walleye in the fall of 1969. Accepting 0.5 ppm set by the Swedish government as a safe level, the Canadian federal government was alarmed by levels of 1 to 2 ppm of mercury found in the walleye flesh. Therefore, in the spring of 1970 the Canada Department of Fisheries and Oceans and the Michigan Department of Natural Resources closed the commercial fishery from southern Lake Huron through to the western basin of Lake Erie. The Lake St. Clair commercial fisheries were substantial, providing about forty small family companies with $1–2 million worth of fish per year.

In response to the mercury crisis, Dow Chemical of Canada voluntarily shut down its mercury cell plants late in 1970 and began using an alternative, non-mercury process to produce chlorine and caustic soda (Hartig 1983). In 1972, a chemical company in Wyandotte, Michigan also stopped using the mercury-based technology and began to use alternative processes.

Dow's Canadian chlor-alkali plant had been in operation for almost twenty years when researchers discovered the environmental contamination caused by its production process. They estimated that it would take a century to rehabilitate a mercury-contaminated system. Ecosystems have, however, proven to be more resilient than initially anticipated. Since mercury discharges into the St. Clair River ended in the early 1970s, concentrations of mercury in Lake St. Clair walleye have declined by over 80 percent (figure

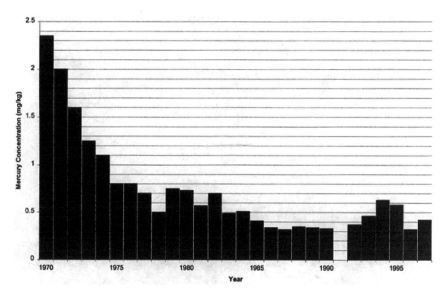

Figure 8.3. Mercury concentrations in 45 cm walleye from Lake St. Clair (data collected by the Ontario Ministry of the Environment; Environment Canada and U.S. Environmental Protection Agency 1999).

8.3). Similar reductions of mercury occurred in other species of fish. The most recent monitoring of fish from Lake St. Clair and the Detroit River has shown that mercury levels in Detroit River fish are slightly lower than in the same species from Lake St. Clair. However, health advisories remain in effect for certain sizes of some fish species from both Lake St. Clair and the Detroit River.

Scientists have attributed this decline of mercury in Lake St. Clair and Detroit River fishes to the end of mercury discharges into the St. Clair River and the movement of mercury-contaminated sediments from Lake St. Clair to Lake Erie. Today, the primary source of mercury is contaminated sediment from historic discharges (Michigan Department of Natural Resources and Ontario Ministry of Environment 1991) and airborne fallout. In the Detroit River, mercury-contaminated sediments can be found in depositional zones, particularly the Trenton Channel. These mercury hot spots are now being evaluated for potential sediment remediation to reduce further the mercury contamination of fish and other organisms inhabiting the Detroit River.

Although the decline of mercury in Lake St. Clair and Detroit River fish is remarkable, this is not the real lesson. It is, rather, that we must prevent future discharges of persistent toxic contaminants such as mercury. The economic and social impacts of the mercury crisis were and continue to be substantial, including the closure of commercial fisheries. We now have the experience and understanding to know that even small amounts of chemicals released into the environment can prove dangerous. We must be vigilant to prevent their release. Experience demonstrates that pollution prevention is much more ecologically sound and cost-effective than environmental remediation.

PCB Contamination

Mercury is not the only contaminant of concern in the Detroit River. PCB contamination is just as problematic. PCBs are a group of synthetic chemicals falling within a larger class of substances known as chlorinated organic compounds. They are chemically stable, fire resistant, do not conduct electricity, and are not volatile at low temperatures. These characteristics made PCBs attractive for a wide range of consumer and industrial uses, such as hydraulic and heat exchange fluid, plasticizers, and as ingredients in caulking compounds, adhesives, paints, printing inks, and carbonless paper. After World War II in particular, PCBs were heavily used worldwide for these purposes (Canadian Council of Resource and Environment Ministers 1986). The chemical stability of PCBs makes them particularly insidious environmental contaminants because once released, they persist, biomagnify, and disperse throughout the food web.

Scientists experimenting with PCBs in labs have discovered a number of health effects associated with exposure to these chemicals that range from disrupted photosynthesis in plankton to reproductive effects in higher animals. In rats and mice, PCBs have been associated with liver cancer. In the 1970s, Great Lakes field researchers found that populations of fish-eating birds, such as herring gulls, suffered reproductive failure and experienced poor adult nesting behavior when exposed to PCBs in the environment. Follow-up tests showed elevated levels of PCBs in both herring gull eggs and adult bird tissue. Similarly, fur farmers using Great Lakes fish with high PCB concentrations as feed suffered severe losses in their mink populations. Subsequent research revealed that PCBs biomagnify in food webs (Canadian Council of Resource and Environment Ministers 1986).

The impact of human ingestion of PCBs is not fully understood, although the immediate effects can be acute. In 1968, for example, some 1,200 people in Japan were exposed to PCBs when a heat exchanger in a rice oil factory leaked the chemical into the rice oil and they consumed the oil

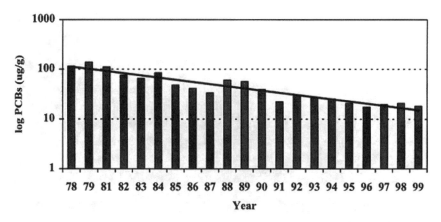

Figure 8.4. PCB levels (1:1 Aroclor 1254:1260) in herring gull eggs from Fighting Island, Detroit River, 1978–99. (These levels display a significant declining temporal trend that has been constant throughout the duration of study; Pekarik and Weseloh 1988.)

over a period of 20 to 190 days. Effects included reproductive problems, hyper-pigmentation, eye discharges, headaches, vomiting, fever, and visual and respiratory problems (Canadian Council of Resource and Environment Ministers 1986). Longer-term effects include an impact on enzyme activity and cancer.

The presence of PCBs in the environment was discovered by accident. In 1966, Swedish scientists searching for the pesticide DDT, a related persistent toxic compound, found PCBs in the fatty tissues of fish and birds. As was the case with the discovery of environmental mercury contamination, finding PCBs in the environment prompted researchers around the world to begin looking for the chemical in their countries. Scientists first identified PCBs in Great Lakes fish in 1970. In 1972, researchers in the Canadian Wildlife Service found PCBs in the eggs of a herring gull colony on Fighting Island in the Detroit River. Although long-term monitoring of PCB concentrations in herring gull eggs collected from Fighting Island shows a declining trend (figure 8.4), researchers are not convinced this is a result of a comparable decline in PCBs in the forage fish consumed by the gulls. Instead, they believe that herring gull diets have recently changed as forage fish become more difficult for the gulls to catch. Recent research into the level of PCBs in the sediment and invertebrate life of the Detroit River demonstrates that PCB levels remain relatively high.

Once it became clear that PCBs were having a significant negative impact on the environment, Monsanto Company, the sole manufacturer of the chemical in North America, began to voluntarily restrict sales. Monsanto limited the sale of PCBs for use in closed systems, such as electrical transformers, where chances of environmental contamination were greatly reduced. This immediately cut PCB use in North America by one-half. In both the United States and Canada PCB production is banned and use is restricted to existing closed systems. In the United States, PCB–based equipment is being replaced in an accelerated phase-out program as rapidly as possible. In Canada, legislation allows PCB electrical equipment to be phased out at the end of its regular service life. Most of this equipment is due to be phased out relatively soon, but some may remain in operation until 2020 (Canadian Council of Resource and Environment Ministers 1986).

Despite the ban on the production and use of PCBs in North America, it is evident that the Detroit River continues to be a major source of these chemicals. Historically, a significant proportion of PCBs was buried in landfills. Contaminated groundwater from these landfills can slowly leach into the river. Today, contaminated sediment is the legacy of historic toxic chemical loadings entering the river. These contaminated sediments are now a source of contamination through biomagnification in the food web. The atmosphere is another significant source.

We must recognize that mercury and PCBs are only two of over fifty chemicals in the Detroit River (see table 8.1). What other chemicals might be present in the Detroit River ecosystem? What can and should be done to prevent such contamination?

Detroit River Management

The United States and Canada have long boasted that their undefended border is an example to the rest of the world of how effectively two independent nations can coexist. In the environmental arena, the establishment of the International Joint Commission (IJC), a binational agency responsible for overseeing how Canada and the United States develop shared water resources, is one example of cooperation. This organization leads joint studies, such as the Upper Great Lakes Connecting Channel Study, which highlighted the state of the Detroit River. Yet there is still a long way to go before the two countries are able to implement a common and effective environmental management plan for the Detroit River. Binational environmental management is hampered by a lack of data and information about the state of the river. A comprehensive and ongoing binational monitoring program is needed, as are common and consistent guidelines for chemicals released to the environment. Finally, a clear understanding of how the river's physical properties affect the chemicals already present in the sediment and how they interact in the environment is necessary for effective binational management of the river.

Environmental monitoring is one of the most important tools for effective management of the earth's resources. A lack of data and information permitted decades of mercury and PCB pollution to occur undetected. The Detroit River was not polluted maliciously; it was done in ignorance. Frequently managers are forced to make decisions with incomplete information and should not be blamed for decisions shown subsequently to be faulty. Ideally, they should have the range of data they require to make timely and accurate environmental management decisions. This is most often provided by long-term monitoring programs.

Detroit River monitoring has an interesting history. After World War II, the Michigan Department of Natural Resources developed a program under which it annually monitored the quality of the river water. In the 1970s, the IJC reviewed the Michigan monitoring plan and included it, with little modification, in the Great Lakes International Surveillance Plan (GLISP). Indeed, the Detroit River monitoring program was among the best in the Great Lakes Basin. Unfortunately, Michigan stopped monitoring in the mid-1980s and only recently has attempted to restore it. At the same time, the federal governments of Canada and the United States also failed to fulfill their obligations under GLISP. Without Canadian and U.S. monitoring, the environmental management of toxic chemicals in the Detroit River returned to the earlier paradigm of management by ignorance.

This is why Detroit River managers remain today unable to answer such questions as: Is the quality of Detroit River water improving? Are there other, new problems we need to address? What are our best remedial options? What will be the benefits of remedial efforts? Unfortunately, through cost-cutting measures the governments ended the monitoring programs necessary to answer these questions, which would have cost a modest eighty cents annually for each person living in the watershed.

Is the river improving? A few remaining monitoring programs appear to indicate that PCB contamination is declining. Canadian Wildlife Service studies of PCBs in herring gulls reveal that concentrations of PCBs are decreasing (see figure 8.4). Biomonitoring programs by the city of Windsor demonstrate that sources of PCBs on the Canadian side are minimal and that they enter the river only during storm events. Unfortunately, there are major landfills on the U.S. side that still contribute to the PCB loads entering the river. Therefore, although levels in gulls have declined, it will be a long time before PCBs are eliminated from the river. PCB monitoring programs, however, allow resource managers to make informed decisions about the environment.

But what of the other 40,000 or so potentially hazardous chemicals that can find their way into the Detroit River ecosystem? Mercury is a good example of management by ignorance. There are no reliable data to estimate how much mercury is still being discharged into the river, a legacy of the failure of governments to maintain GLISP. In addition, Canadian and U.S. data cannot be compared as the quality assurance protocols established under GLISP have also been abandoned.

High PCB and mercury levels mean that both countries still have fish consumption advisories, yet there is little information about what can be done to restore the river to a state where the fish can be eaten. Millions of dollars have been and continue to be spent removing PCBs and other contaminants from the river (table 8.2), but has this contaminated sediment remediation moved us forward? If so, why is the river still considered one of the most polluted in North America? With so little monitoring it is difficult to determine if the classification is justified, or if it is a legacy of the river's reputation.

Another barrier to effective management of the Detroit River ecosystem is the lack of consistency among governments with management responsibility. Guidelines establishing acceptable levels of chemical contamination in water and food are quite different in Canada and the United States. Developing common and consistent guidelines is an example of the kind of stronger cross-border linkages required to prevent, reduce, and control toxic contamination of the river. A coordinated, binational management plan requires common preventive and remedial strategies and commonly acceptable levels of discharges to ensure effective remedial efforts.

An important tool in the joint management effort is the development of common models to help decision makers establish priorities, identify the best means to solve problems, and predict the results of various remedial options. Confusion can easily arise from the variety of fish and animals in the river potentially affected by toxic chemicals, the range of different chemicals, and the difficulty of dealing with historic and current chemical loadings. It is, therefore, essential to develop accurate models.

Many types of models are available. Some examine the way known chemical contaminants are transported in the river and what influences

TABLE 8.2
Sediment Remediation Projects in the Detroit River Watershed (Michigan)

River	Sediment Remediation Project	Date	Cost (millions of $)	Estimated Volume of Sediment Removed
Rouge	Evans Products Ditch Site	1997	0.75	7,300 m³
Rouge	Newburgh Lake	1997–98	11.0	306,000 m³
Detroit	Carter Industrial Site	1986–87 residential 1995–96 soil excavation	7.0	35,132 m³
Detroit	Elizabeth Park Marina Wayne County	1993	1.3	3,100 m³
Detroit	Monguagon Creek	1997	3.0	19,300 m³
Detroit	Conners Creek	2002–2003	To be determined	114,000 m³
Detroit	Black Lagoon	Planned for 2004	9.0	30,000 m³
Huron	Willow Run Creek	1998	70.0	336,400 m³
River	Ford Motor Company Site	1996–1997	6.0	20,000 m³

where they stop. Other models focus on chemical biomagnification in fish and wildlife. Still others predict the hazard of unknown or untested chemicals. Such models are important because only about twenty of the chemicals listed in table 8.1 have been sufficiently tested to develop guidelines for acceptable levels to protect human and environmental health. These models will be important tools in deciding what to do about contaminants in the Detroit River.

The Detroit River is physically complex and contaminated with a variety of chemicals. Many other chemicals may remain undetected, which means that insects, fish, birds, other animals, and humans are exposed to chemical mixtures. Understanding how these chemicals interact is, therefore, very important and requires not only good monitoring data but good research tools to evaluate the hazard of chemicals in the river. The exchange and acquisition of data and information on the World Wide Web offers a significant advantage. Web pages are being developed to give researchers access to Detroit River information. The effects of accidental spills can be rapidly assessed and the benefits of remediation can be identified and measured using common tools.

Conclusion

What will we do about contaminants in the Detroit River? What can citizens do to help? There is an important remedial effort underway in the Detroit River watershed. People who use, enjoy, or affect all aspects of the river have come together to develop the Detroit River Remedial Action Plan. It is important for everyone with an interest in the quality of the river to get involved with this and other cleanup efforts. Learn more about the river, take a river tour, and participate in a cleanup activity. Participate in public meetings, especially where the government is soliciting advice to develop management priorities. Urge your Member of Parliament or congressional representative to support funding for Detroit River remediation. Let them know you think this is an important effort.

Postwar society had ignorance as an excuse for not recognizing the problem of contaminants in the river and failing to do something about the situation in time to prevent the contaminants from spreading. Today ignorance is not an excuse. We need to clean up the river and reverse its reputation as one of the most polluted rivers in North America. Many people living along the river no longer trust it as a source of drinking water, many do not eat the fish, and others refuse to use the river for recreational activities. When people stop using a resource it is difficult to protect and conserve it. This is particularly true of the Detroit River. Its poor reputation does not promote tourism, nor is it particularly attractive to new businesses, which contribute to the economic health of the region. Ignoring the river's reputation is a mistake that neither Americans nor Canadians can afford to make. The river affects the lives and future of everyone living in the watershed.

Notes

The authors would like to acknowledge the help of Sara Gewurtz, Marcia Valiante, Chip Weseloh, John Hartig, and Ed Kimball.

1. Bioaccumulation is a process where chemicals are usually retained in fatty body tissue and increase in concentration over time. Biomagnification is the increase in concentration of accumulated toxics in species higher in the food web as contaminated food species are consumed.
2. Parts per million (ppm) is a common scientific term of measurement for describing the concentration of a particular element or compound in air, water, or sediment. It is equivalent to mg/kg. One part per million is comparatively equal to a peanut in relation to an elephant.

References

Canadian Council of Resource and Environment Ministers. 1986. *The PCB Story.* Toronto: The Council.

Carr, R. L., C. E. Finsterwalder, and M. J. Schibi. 1972. "Chemical Residues in Lake Erie Fish: 1970–71." *Pesticide Monitoring Journal* 6:23–26.

Carson, Rachel. 1962. *Silent Spring.* Boston: Houghton Mifflin.

Ciborowski, J.J.H., and L. D. Corkum. 1988. "Organic Contaminants in Adult Aquatic Insects of the St. Clair and Detroit Rivers, Ontario, Canada." *Journal of Great Lakes Research* 14:148–56.

D'Itri, F. M. 1972. *The Environmental Mercury Problem.* Cleveland: Chemical Rubber Company.

Environment Canada and U.S. Environmental Protection Agency. 1999. *State of the Great Lakes.* Burlington, Ont., and Chicago: Environment Canada and U.S. EPA

Evans, R. J., J. D. Bails, and F. M. D'Itri. 1972. "Mercury Levels in Muscle Tissues of Preserved Museum Fish." *Environmental Science and Technology* 6:901–5.

Fallon, M. E., and F. J. Horvath. 1985. "Preliminary Assessment of Contaminants in Soft Sediments of the Detroit River." *Journal of Great Lakes Research* 11:373–78.

Fimreite, N., W. H. Holsworth, J. A. Keith, P. A. Pearce, and I. M. Gruchy. 1971. "Mercury in Fish and Fish-eating Birds Near Sites of Industrial Contamination in Canada." *Canadian Field-Naturalist* 85:211–20.

Froese, K. L., D. A. Verbrugge, S. A. Snyder, F. Tilton, M. Tuchman, A. Ostaszewski, and J. P. Giesy. 1997. "PCBs in the Detroit River Water Column." *Journal of Great Lakes Research* 23:440–49.

Gewurtz, S. B., R. Lazar, and G. D. Haffner. Unpublished manuscript. "Biomonitoring of Bioavailable PAH and PCB Water Concentrations in the Detroit River Using the Freshwater Mussel, *Elliptio complanata.*"

Hamdy, Y., and L. Post. 1985. "Distribution of Mercury, Trace Organics, and Other Heavy Metals in Detroit River Sediments." *Journal of Great Lakes Research* 11:353–65.

Hartig, John. 1983. "Lake St. Clair: Since the Mercury Crisis." *Water Spectrum* 15(1):18–25.

Hess, J. L., and E. D. Evans. 1972. *Heavy Metals in Surface Waters, Sediments and Fish in Michigan.* Michigan Water Resources Commission, Department of Natural Resources, Lansing.

Holdrinet, F. M., H. E. Braun, R. L. Thomas, A.L.W. Kemp, and J. M. Jacquet. 1977. "Organochlorine Insecticides and PCBs in Sediments of Lake St. Clair (1970 and 1974) and Lake Erie (1971)." *Science of the Total Environment* 8:205–27.

Kaiser, K.L.E., M. E. Comba, H. Hunter, R. J. Maguire, R. J. Tkacz, and R. F. Platford. 1985. "Trace Organic Contaminants in the Detroit River." *Journal of Great Lakes Research* 11:368–99.

Metcalfe, T. L., C. D. Metcalfe, E. R. Bennett, and G. D. Haffner. 2000. "Distribution of Toxic Organic Contaminants in Water and Sediment in the Detroit River." *Journal of Great Lakes Research* 26:55–64.

Michigan Department of Natural Resources and Ontario Ministry of Environment. 1991. *Stage 1 Remedial Action Plan for the Detroit River.* Lansing and Sarnia, Ont.

Lear, L. 1997. *Rachel Carson: Witness for Nature.* New York: Henry Holt.

Pekarik, C., and D. V. Weseloh. 1988. "Organochlorine Contaminants in Herring Gull Eggs from the Great Lakes, 1974–1995: Change Point Regression and Short-Term Regression." *Environmental Monitoring and Assessment* 53:77–115.

Platford, R. F., R. J. Maguire, R. J. Tkacz, M. E. Comba, and K. L. E. Kaiser. 1985. "Distribution of Hydrocarbons and Chlorinated Hydrocarbons in Various Phases of the Detroit River." *Journal of Great Lakes Research* 11:379–85.

Reinert, R. E. 1970. "Pesticide Concentrations in Great Lakes Fish." *Pesticide Monitoring Journal* 3:233–40.

Reinke, J., J. F. Uthe, and D. Jamieson. 1972. "Organochlorine Pesticides Residues in Commercially Caught Fish in Canada—1970." *Pesticide Monitoring Journal* 6:43–49.

Russell, R. W., F.A.P.C. Gobas, and G. D. Haffner. 1999. "Role of Chemical and Ecological Factors in Trophic Transfer of Organic Chemicals in Aquatic Food Webs." *Environmental Toxicological Chemicals* 18:1250–57.

Weseloh, D. V., P. Mineau, and J. Struger. 1990. "Geographical Distribution of Contaminants and Productivity Measures of Herring Gulls in the Great Lakes: Lake Erie and Connecting Channels 1978/79. *Science of the Total Environment* 91:141–59.

Lessons from Sentinel Invertebrates: Mayflies and Other Species

Jan J. H. Ciborowski

A quick look at the bottom of the Detroit River may not be a very good indicator of what is really there. We might recognize a zebra mussel attached to a rock, and a snail or two. But a wide variety of animals can be found hidden under the stones and buried in the mud. Farrara and Burt (1993) report finding 176 different types of invertebrates in a 1980 survey of sediments of the river. As a group, we call these organisms "zoobenthos" (zoo = animal; benthos = living on the bottom) or "benthic invertebrates." They play an important role in the functioning of the ecosystem because they are the key link between algae and organic matter that form the base of the food web, and the amphibians, fishes, and waterfowl higher in the ecosystem. Like earthworms and insects on land, zoobenthos convert living and dead plant materials into animal protein.

Because the Detroit River is large and complex, it offers a full range of aquatic habitats that provide environmental conditions suitable for many types of biota. As the gateway between the lower and upper Great Lakes, the Detroit River provides passage for ships from around the world. Consequently, it has become colonized by all sorts of exotic aquatic invertebrates. Indeed, zebra mussels were first collected in Lake St. Clair near the head of the Detroit River (Hebert, Muncaster, and Mackie 1989). The European spiny amphipod *(Echinogammarus ischnus)* was first found near Amherstburg (Witt, Hebert, and Morton 1997). As well, the Detroit River has produced adult mitten crabs (natives of east Asia; Nepszy and Leach 1973; pers. obs.), and the tiny colonial jellyfish *Cordylophora caspia* (Wood 2003). Other introduced zoobenthos may also be lurking in the river, awaiting discovery.

Habitats and Their Residents

The abundance of the different types of aquatic invertebrates in various places of the river is controlled largely by the current and its effects on the river bottom. These differences in habitat must be taken into account when one attempts to evaluate the potential effects of human-related activities on the benthic community.

Erosional Zones

Currents in the main part of the river wash away fine sediments, leaving a substrate composed mainly of stones and hard clay (erosional areas). Some animals can shelter beneath or between the stones, but the most abundant species can attach themselves to the substrate or build shelters.

Exposed surfaces of the substrate develop growths of algae, which are fed upon by grazing animals such as freshwater limpets. Other animals (filter-gatherers) collect algae, zooplankton, and bits of organic debris from the water as it flows past them (see table 9.1).

Bivalves have historically been important filter-gatherers in the Detroit River. Sandy and gravelly sediments supported high densities and diversity

TABLE 9.1

The Most Common Types of Zoobenthos in Different Habitats of the Detroit River

Erosional Zones

Dreissena (Zebra mussels)
Cheumatopsyche (Net-spinning caddisflies)
Hydra
Dugesia (flatworms)
Amnicola, Ferrissia, and other snails
Gammarus (amphipod crustaceans)
Pisidium (fingernail clams)

Depositional Zones

Tubificidae (worms)
Manayunkia (worms)
Chironomidae (midges)
Hexagenia (mayflies)

Sources: Hudson et al. 1986; Farrara and Burt 1993; Wood 2003.

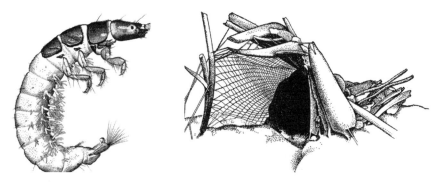

Figure 9.1. Left: *Cheumatopsyche*, the most common caddisfly in the Detroit River; right: retreat and capture net of *Hydropsyche* (Wiggins 1977).

of unionid mussels and fingernail clams (Hudson et al. 1986). Zebra mussels became established in the river in the late 1980s (Hebert, Muncaster, and Mackie 1989; Yankovich and Haffner 1993). Zebra mussels' ability to secrete sticky "byssus" threads permitted them to quickly colonize virtually all hard-substrate regions of the river. By 1990, zebra mussels were observed at over 70 percent of the stations sampled (Farrara and Burt 1993). They are now found almost everywhere in the river that attachment sites are available (Wood 2003).

The small spaces created among the bodies of tightly packed zebra mussel colonies provide a sheltered habitat for other types of animals. Amphipods (small shrimp-like crustaceans) are often associated with zebra mussels. These scavengers feed upon zebra mussel feces and the river debris that collects in the dead-water zones around the mussels.

The caterpillar-like larvae of net-spinning caddisflies are also abundant and effective filter-gatherers. They are able to spin silk, which they use to glue together bits of stones and debris into a shelter. Other silk is spun into nets (figures 9.1 and 9.2). The larva feeds upon the catch of the day from among items that become caught in the net. Net-spinning is a complicated activity. Researchers in Sweden have suggested that net-spinning caddisfly behavior is a good indicator of potential toxic effects of effluents. Petersen and Petersen (1984) found that caddisflies living in water downstream of pulp and paper mills were much more likely to make mistakes during net production (irregularities in the meshes) than caddisflies of the same species collected from reference streams. Caddisfly nets have not been examined for aberrations in the Detroit River. Davis, Hudson, and Armitage (1991)

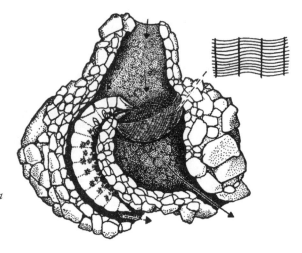

Figure 9.2. *Macronema zebratum* net-spinning caddisfly in its retreat (Wiggins 1977).

describe the distribution and species of adult caddisflies collected from the shores of the Detroit River.

Other Detroit River animals that feed on particles carried by the water include hydrozoans, which are tiny freshwater cousins to jellyfish. The well-known *Hydra* is abundant in the deeper, rapid parts of the river. The colonial hydrozoan *Cordylophora caspia* has also become established in the river (Wood 2003). This animal, probably introduced from the Black Sea (the same home as zebra mussels), forms colonies of individuals, each of which catches water fleas and other small plankton with its stinging tentacles (figure 9.3). Nutrients from the digested food are then shared among members of the colony. A single specimen of *Cordylophora* was collected during a 1983 survey of the Detroit River (Hudson et al. 1986). Recent surveys have shown that *Cordylophora* has become more abundant and widespread (Wood 2003).

Depositional Zones

Slower-flowing parts of the Detroit River (depositional areas) have a muddy or sandy bottom. These are also the areas where debris carried from upstream settles out. The benthic animals living here burrow into the mud and feed upon the organic debris and the attached bacteria and fungi present in the sediments (table 9.1). When the organic content of the sediments is high, bacterial respiration can remove much of the oxygen from the water, making the habitat suitable only for tolerant species. These are regions where pollutants that are not water-soluble tend to collect. Because oils and trace amounts of metals adhere to the organic matter in the mud, the zoobenthos of depositional zones tend to bioaccumulate these materials from their food.

Figure 9.3.
Cordylophora, a colonial hydrozoan from the Detroit River. Inset in upper lefthand corner shows a feeding polyp. Actual width of the polyp is about 1 mm. (J. Ciborowski, University of Windsor).

The most common benthic animals living in depositional zones are worms, midge larvae, and *Hexagenia* mayfly nymphs. Each of these types of animals has adaptations that help it acquire enough oxygen from the water. Mayfly nymphs dig U-shaped burrows (figure 9.4). They undulate their abdomen and wave their feather-like gills, which forces oxygen-rich water through the burrow. Because mayflies cannot survive in water that is poor in oxygen, they are good indicators of the amount of organic pollution (e.g., sewage). In contrast, worms and some types of midges are able to thrive under poor aquatic conditions because their blood contains hemoglobin, which makes them efficient at extracting oxygen even when there is very little in the water. Mud that is rich in bacteria and organic debris can sustain enormous numbers of worms and midges. However, high concentrations of pollutants such as heavy metals or industrial chemicals are toxic to all biota.

Some reaches of the Detroit River develop beds of aquatic plants, which provide a rich, three-dimensional habitat for zoobenthos (Hudson et

Figure 9.4. Nymph of *Hexagenia bilineata* in artificial burrow (Fremling 1970).

al. 1986; Manny, Edsall, and Jaworski 1988). Aquatic plants contain compounds that make them unpalatable to many invertebrates. But single-celled algae growing on the leaves are grazed by snails and other invertebrates. The dense vegetation provides shelter for the invertebrates from fishes, but also serves as a lair for predatory insects. Weed beds probably have as wide a variety of invertebrates as any other habitat in the Detroit River, but they have seldom been studied.

Using Zoobenthos Distribution to Assess Local Conditions

By evaluating the most common types of animals, we can infer local environmental conditions. For example, when water quality conditions are good, one expects to find 100 *Hexagenia* larvae/m^2 or more in clean muddy sediments of Lake Erie (Wright and Tidd 1933). Areas dominated by worms and midges, rather than mayflies, are classed as having degraded water quality or benthic conditions. Very high densities of worms (over 5,000/m^2; Wright and Tidd 1933) and midges suggest organic enrichment (from sewage, or nutrient inputs from agriculture), whereas very low densities of worms and midges (and other invertebrates) in severely polluted areas imply that contamination of the sediments with metals or chemicals is a cause of degradation.

Some of the earliest Great Lakes zoobenthos surveys were conducted in 1929–30 by Wright and Tidd (1933) in western Lake Erie and at the mouth of the Detroit River, as well as other Lake Erie tributaries. Among other zoobenthos, they reported snails, fingernail clams, and worms. Conspicuously absent were mayfly larvae, indicating light to moderate pollution at the Detroit River mouth. At that time, the most seriously polluted rivers were those most heavily industrialized—the Maumee River and the River Raisin. However, even then, siltation due to erosion from farming areas in Michigan and Ohio was recognized as a significant cause of river-bottom degradation (Wright and Tidd 1933).

The Trenton Channel has long been identified as a degraded area based on zoobenthos composition and abundance. Wright and Tidd (1933) noted a paucity of live fingernail clams in the channel, reflecting degraded water quality. Surveys conducted between 1949 and 1956 showed that the lower Detroit River and the western Trenton Channel were dominated by pollution-tolerant forms, indicating a decrease in water quality from the 1929–30 surveys. Carr and Hiltunen (1965) showed that the spatial extent and severity of degradation at the mouth of the Detroit River had increased substantially from that described by Wright and Tidd (1933). Later surveys reported that although the river mouth contained only very pollution-tolerant organisms (worms and leeches), zoobenthos composition and abundance upstream of Belle Isle was indicative of good water quality (Vaughan and Harlow 1965).

Several surveys of the bottom fauna of the Detroit River were conducted between the 1960s and 1990 (Thornley 1985; Hudson et al. 1986; Farrara and Burt 1993). In 1968, the bottom fauna over large tracts of the river suggested that sediments and water quality were degraded. Mayflies were found in only about 25 percent of the locations sampled, and then only in low numbers (10–20/m²; Thornley and Hamdy 1984). Mayflies were completely absent from the U.S. shoreline except near the upstream end of Belle Isle. Almost no zoobenthos could be found in the vicinity of Zug Island. Downstream, worm densities rose to 500,000 per m². Immediately downstream of the confluence of the Rouge and Detroit Rivers, pollution tolerant worms numbered over 1,000,000 per m² in both a 1968 survey and a 1980 survey, indicating long-term, severe, organic enrichment (Thornley and Hamdy 1984). Although worms were almost the only organisms present downstream along the U.S. shoreline, densities declined in the Trenton Channel, implying toxic effects.

Pollution controls put in place during the 1970s resulted in improved water and sediment quality in many areas. When the river was surveyed again in 1980, mayflies were found at over 70 percent of the locations examined, and they were five times more abundant than they had been in 1968 (Thornley 1985).

Few changes in either the distribution or abundance of mayfly nymphs were seen between the 1980 survey, a 1983 investigation (Hudson et al. 1986), and a study done in 1990 (Farrara and Burt 1993). In 1990, *Hexagenia* mayflies were found at about 60 percent of locations sampled, at densities of between 8 and 100 nymphs/m². Worms and midges remained the most common invertebrates along the U.S. shoreline of the river downstream from Zug Island. Little benthic sampling was conducted in the Detroit River through the 1990s, so information on the health of the zoobenthic community during this time is scarce. However, the flying adult stages of aquatic insects became more numerous along both the Canadian and U.S. sides of the river (Ciborowski and Corkum 1988; Corkum, Ciborowski, and Lazar 1997), suggesting that some improvements in river condition have been occurring.

The Detroit River Remedial Action Plan (RAP) has identified degraded zoobenthos as one of nine impaired beneficial uses (Michigan Department of Natural Resources and Ontario Ministry of Environment 1991). Regarding the degradation of zoobenthos, the Detroit River RAP has concluded the following:

> Surveys of the benthic communities of the Detroit River indicate that although there have been significant improvements since the 1960s, the zoobenthic invertebrate communities are severely degraded along Michigan's Trenton Channel. Pollution tolerant organisms indicative of extreme organic enrichment comprised 95% of the benthic invertebrates collected in this reach in 1985–1986. Mid-river benthic communities were considerably better than those found along the Michigan shoreline, however communities were depressed in the depositional zones of the navigation channels. A balanced benthic community structure indicative of satisfactory water quality conditions was noted along the Ontario shoreline. Canadian inputs to the Detroit River did not severely disrupt the macrozoobenthos. . . . Improvements in the benthic communities were noted in 1980 in the entire river and especially along the Michigan shoreline from Belle Isle to the mouth of the Rouge River and along most of the Ontario shoreline. (5–6)

Giesy et al. (1988) confirmed that the sediments themselves were toxic to zoobenthos by showing reduced survival of midge larvae reared in sediments collected from the Trenton Channel. The most recent analyses of zoobenthic abundance and community composition, performed in 1999, indicate that apart from the appearance of several exotic species, relatively little change in the distribution, composition, or abundance of zoobenthos has occurred through the 1990s (Wood 2003).

Benthic Invertebrates and Contaminants

Zoobenthos can be effective sentinels of environmental quality as conditions improve to the point where they can survive and grow to maturity. Aquatic invertebrates that live in or on moderately contaminated sediments can bioaccumulate the compounds within their own bodies. Some types of invertebrates are able to break down the pollutants, whereas the body burden of contaminants in other organisms gradually increases through the course of their lifetimes. Midge larvae can metabolize polychlorinated aromatic hydrocarbons (PAHs) (Harkey, Landrum and Klaine 1994) and other organic contaminants. Mayfly nymphs and mussels simply store many types of material, especially those that are very insoluble in water (Landrum and Poore 1988). The tissues of these invertebrates often have much higher concentrations of polychlorinated biphenyls (PCBs) and pesticides than the sediments in which they dwell (Gobas et al. 1989; Drouillard et al. 1996). This makes them useful biomonitoring indicators of local conditions. Kauss and Hamdy (1985), Muncaster et al. (1989), Muncaster, Hebert, and Lazar (1990), and others caged *Elliptio* mussels in the Detroit and St. Clair Rivers to generate estimates of water concentrations of PCBs and chlorinated hydrocarbons. The filter-feeding activities of zebra mussels results in their especially rapid uptake of contaminants. Therefore, they have been used by several workers as biomonitors of the local distribution and levels of these compounds in the Detroit River (Morrison et al. 1995).

The metabolized breakdown products of organic contaminants ingested by biota are sometimes more toxic than the original chemicals. Consequently, organisms capable of metabolizing compounds such as PAHs and PCBs can suffer developmental abnormalities. The crossed bills and clubbed feet occasionally reported in colonial water birds nesting in highly contaminated regions of the Great Lakes are thought to be caused by such effects (Giesy, Ludwig, and Tillitt 1994). Zoobenthos are also susceptible to contaminant-induced deformities (Hamilton and Saether 1971; Vuori and Parkko 1996). Pollution-tolerant midge larvae from the Detroit River have been

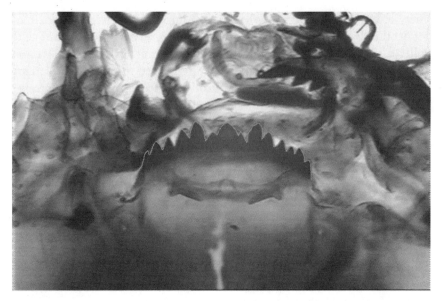

Figure 9.5. Micrograph of the ventral side of the head of a midge larva, *Chironomus*. The mentum (row of teeth in the middle of the photo) normally has three teeth in the center and six teeth on either side. This deformed individual has only five teeth on the left side of its mentum (actual width of the head is about 2 mm) (J. Ciborowski, University of Windsor).

examined for indications of elevated incidences of mouthpart deformities (extra or missing teeth on the mentum; see figure 9.5).

Hudson and Ciborowski (1996) found that deformities were much more common in midges collected from sites at Peche Island, Turkey Island, and the Trenton Channel than in midges from reference sites in Anchor Bay (Lake St. Clair) and a riverside pond. Doherty et al. (1999) compared incidences of deformities in midges collected by the U.S. Geological Survey from the mouth of the Detroit River in 1982 and 1993. In 1982, over 6 percent of larvae were deformed. But by 1993, the level of deformities had been reduced by half, suggesting that sediment quality had improved over that interval.

Improvements in Detroit River water quality have been most obviously shown in the numbers of night-flying insects that are attracted to streetlights and storefronts along the river during warm summer evenings. Both *Hexagenia* mayflies (also called junebugs or fishflies) and moth-like caddisflies emerge from the river in summer to mate and lay their eggs. Caddisflies can be found on the wing throughout the summer. *Hexagenia* mayflies are most abundant for a few weeks from the end of May until the middle of June. Strong winds can carry the insects long distances inland from the river, but typically most travel only a few hundred meters (Kovats, Ciborowski, and Corkum 1996). Although the insects are a nuisance, they are an important food for birds and fishes during their emergence period. They also provide us with another valuable tool for monitoring contaminant levels in the river. On a warm evening, a black light placed beside the river will quickly attract enough biomass to provide a sample that can be analyzed for PCBs, heavy metals, and other pollutants associated with contaminated sediments (Corkum, Ciborowski, and Kovats 1995). Ciborowski and Corkum (1988), Kovats (1990), and Corkum, Ciborowski, and Lazar (1997) analyzed organic contaminant burdens in *Hexagenia* mayflies emerging at the head of the Detroit River near Peche Island. Concentrations of PCBs, pesticides, and other organochlorine compounds were virtually identical in 1986, 1989, and 1994. However, the numbers of emerging insects, and their distribution along the river, continued to increase through the 1990s (Corkum, Ciborowski, and Lazar 1997).

Summary

The benthic community of the Detroit River can provide a unique perspective on the river's condition. Water quality improvements through the 1970s clearly had a beneficial effect on the zoobenthos, which seemed to be fairly constant throughout the 1980s. A lack of research funding has limited our knowledge of changes occurring in the 1990s. Increasing numbers of emergent insects suggest that the overall quality of the river is continuing to improve. However, the elevated burdens of contaminants observed in zoobenthos and the commonness of deformed midge larvae at several locations in the river indicate that there are still considerable challenges to be addressed.

LITERATURE CITED

Carr, J. F., and J. K. Hiltunen. 1965. "Changes in the Bottom Fauna of Western Lake Erie from 1930 to 1961." *Limnology and Oceanography* 10:551–69.

Ciborowski, J. J. H., and L. D. Corkum. 1988. "Organic Contaminants in Adult Aquatic Insects of the St. Clair and Detroit Rivers, Ontario, Canada." *Journal of Great Lakes Research* 14:148–56.

Corkum, L. D., J. J. H. Ciborowski, and Z. E. Kovats. 1995. "Aquatic Insects as Biomonitors of Ecosystem Health in the Great Lakes Area of Concern." In *Biomonitors and Biomarkers as Indicators of Environmental Change: A Handbook,* ed. F. M.

Butterworth, L. D. Corkum, and L. J. Guzmán-Rincón, 31–44. New York: Plenum.

Corkum, L. D., J. J. H. Ciborowski, and R. Lazar. 1997. "The Distribution and Contaminant Burdens of Adults of the Burrowing Mayfly, *Hexagenia,* in Lake Erie." *Journal of Great Lakes Research* 23:383–90.

Davis, B. A., P. L. Hudson, and B. J. Armitage. 1991. "Distribution and Abundance of Caddisflies (Trichoptera) in the St. Clair-Detroit River System." *Journal of Great Lakes Research* 17:522–35.

Doherty, M. S. E., L. A. Hudson, J. J. H. Ciborowski, and D. W. Schloesser. 1999. "Morphological Deformities in Larval Chironomidae (Diptera) from the Western Basin of Lake Erie: A Historical Comparison." *In Proceedings of the 25th Annual Aquatic Toxicity Workshop: October 18–21, 1998, Quebec City,* ed. R. Van Coillie, R. Chasse, L. Hare, C. Julien, L. Martel, C. Thellen, and A. J. Niimi, 98. *Can. Tech. Rep. Fish. Aquat. Sci. No. 2260: 134 p.*

Drouillard, K. G., J. J. H. Ciborowski, R. Lazar, and G. D. Haffner. 1996. "Estimation of the Uptake of Oganochlorines by the Mayfly *Hexagenia limbata* (Ephemeroptera: Ephemeridae)." *Journal of Great Lakes Research* 22:26–35.

Farrara, D. G., and A. J. Burt. 1993. *Environmental Assessment of Detroit River Sediments and Benthic Macroinvertebrate Communities–1991.* Vol. 1. Prepared for Ontario Ministry of the Environment, Toronto.

Fremling, C. R. 1970. *Mayfly Distribution as a Water Quality Index.* U.S. Environmental Protection Agency, Water Pollution Control Research Series, Report No. 16030 DQH 11/70, Washington, DC.

Giesy, J. P., R. L. Graney, J. L. Newsted, C. J. Rosiu, A. Benda, R. G. Kreis, Jr., and F. J. Horvath. 1988. "Comparison of Three Sediment Bioassay Methods Using Detroit River Sediments." *Environmental Toxicology and Chemistry* 7:483–98.

Giesy, J. P., J. P. Ludwig, and D. E. Tillitt. 1994. "Deformities in Birds of the Great Lakes Region: Assigning Causality." *Environmental Science and Technology* 28:128A–135A.

Gobas, F. A. P. C., D. C. Bedard, J.J.H. Ciborowski, and G. D. Haffner. 1989. "Bioaccumulation of Chlorinated Hydrocarbons by the Mayfly *(Hexagenia limbata)* in Lake St. Clair." *Journal of Great Lakes Research* 15:581–88.

Hamilton, A. L., and O. E. Saether. 1971. "The Occurrence of Characteristic Deformities in the Chironomid Larvae of Several Canadian Lakes." *Canadian Entomologist* 103:353–68.

Harkey, G. A, P. F. Landrum, and S. J. Klaine. 1994. "Comparison of Whole-sediment, Elutriate and Pore-water Exposures for Use in Assessing Sediment-associated Organic Contaminants in Bioassays." *Environmental Toxicology and Chemistry* 13:1315–29.

Hebert, P. D. N., B. W. Muncaster, and G. L. Mackie. 1989. "Ecological and Genetic Studies of *Dreissena polymorpha* (Pallas): A New Mollusc in the Great Lakes." *Canadian Journal of Fisheries and Aquatic Sciences* 46:1587–91.

Hudson, L. A., and J. J. H. Ciborowski. 1996. "Spatial and Taxonomic Variation in Incidence of Mouthpart Deformities in Midge Larvae (Diptera: Chironomidae)." *Canadian Journal of Fisheries and Aquatic Sciences* 53:297–304.

Hudson, P. L., B. M. Davis, S. J. Nichols, and C. M. Tomcko. 1986. "Environmental

Studies of Macrozoobenthos, Aquatic Macrophytes, and Juvenile Fishes in the St. Clair–Detroit River System, 1983–1984." U.S. Fish and Wildlife Service, Great Lakes Fish Lab, Administrative Report No. 86-7, Ann Arbor, MI.

Kauss, P. B., and Y. Hamdy. 1985. "Biological Monitoring of Organochlorine Contaminants in the St. Clair and Detroit Rivers Using Introduced Clams." *Journal of Great Lakes Research* 11:247–63.

Kovats, Z. E. 1990. "Adult Aquatic Insects as Biomonitors of Organochlorine Contamination in Freshwater Habitats." Master's thesis, University of Windsor, Ontario.

Kovats, Z. E., J.J.H. Ciborowski, and L. D. Corkum. 1996. "Inland Dispersal of Adult Aquatic Insects." *Freshwater Biology* 36:265–76.

Landrum, P. F., and R. Poore. 1988. "Toxicokinetics of Selected Xenobiotics in *Hexagenia Limbata*." *Journal of Great Lakes Research* 14:427–37.

Manny, B. A., T. A. Edsall, and E. Jaworski. 1988. *The Detroit River, Michigan: An Ecological Profile.* U.S. Fish and Wildlife Service Biological Report 85(7.17), Ann Arbor, MI.

Michigan Department of Natural Resources and Ontario Ministry of Environment. 1991. *Detroit River Remedial Action Plan: Stage 1.* Lansing, MI and Sarnia, Ont.

Morrison H., T. Yankovich, R. Lazar, and G. D. Haffner. 1995. "Elimination Rate Constants of 36 PCBs in Zebra Mussels (*Dreissena polymorpha*) and Exposure Dynamics in the Lake St. Clair-Lake Erie Corridor." *Canadian Journal of Fisheries and Aquatic Sciences* 52:2574–82.

Muncaster, B. W., P. D. N. Hebert, and R. Lazar. 1990. "Biological and Physical Factors Affecting the Body Burden of Organic Contaminants in Freshwater Mussels." *Archives of Environmental Contamination and Toxicology* 19:25–34.

Muncaster, B. W., D. J. Innes, P. D. N. Hebert, and G. D. Haffner. 1989. "Patterns of Organic Contaminant Accumulation by Freshwater Mussels in the St. Clair River, Ontario." *Journal of Great Lakes Research* 15:645–53.

Nepszy, S. J., and J. H. Leach. 1973. "First Records of the Chinese Mitten Crab, *Eriocheir sinensis* (Crustacea: Brachyura) from North America." *Journal of the Fisheries Research Board of Canada* 30:1909–10.

Petersen, L. B. M., and R. C. Petersen, Jr. 1984. "Effect of Kraft Pulp Mill Effluent and 4,5,6 Trichloroguaiacol on the Net Spinning Behavior of *Hydropsyche angustipennis* (Trichoptera)." *Ecological Bulletins* 36:68–74.

Thornley, S. 1985. "Macrozoobenthos of the Detroit and St .Clair Rivers with Comparison to Neighboring Waters." *Journal of Great Lakes Research* 11:290–96.

Thornley, S., and Y. Hamdy. 1984. *An Assessment of Bottom Fauna and Sediments of the Detroit River.* Ontario Ministry of Environment Report, Toronto.

Vaughan, R. D., and G. L. Harlow. 1965. *Report on Pollution of Detroit River, Michigan Waters of Lake Erie, and Their Tributaries.* U.S. Department of Health, Education, and Welfare, Public Health Service, Division of Water Supply and Pollution Control. Washington, DC: U.S. GPO.

Vuori, K. M., and M. Parkko. 1996. "Assessing Pollution of the River Kymijki via Hydropsychid Caddisflies: Population Age Structure, Microdistribution and Gill Abnormalities in the *Cheumatopsyche lepida* and Hydropsyche pellucidula Larvae." *Archives für Hydrobiologie* 136:171–90.

Wiggins, G. B. 1977. *Larvae of the North American Caddisfly Genera (Trichoptera).* Toronto: University of Toronto Press.

Witt, J. D. S., P. D. N. Hebert, and W. B. Morton. 1997. "*Echinogammarus ischnus:* Another Crustacean Invader in the Laurentian Great Lakes Basin." *Canadian Journal of Fisheries and Aquatic Sciences* 54:264–68.

Wood, S. 2003. "The Use of Benthic Community Composition as a Measure of Contaminant Induced Stress in the Sediments of the Detroit River." Master's thesis, University of Windsor, Ontario.

Wright, S., and W. M. Tidd. 1933. "Summary of Limnological Investigations in Western Lake Erie in 1929 and 1930." *Transactions of the American Fisheries Society* 63:271–85.

Yankovich, T. L., and G. D. Haffner. 1993. "Habitat Selectivity by the Zebra Mussel *(Dreissena polymorpha)* on Artificial Substrates in the Detroit River." In *Zebra Mussels: Ecology, Impacts, and Control,* ed. T. F. Nalepa and D. W. Schloesser, 175–81. Ann Arbor, MI: Lewis Publishers.

SETTING PRIORITIES FOR CONSERVING AND REHABILITATING DETROIT RIVER HABITATS

Bruce A. Manny

We all have a mental picture of our own habitat. It's where we live, work, and play. Here, I discuss habitat for wild animals and plants in the Detroit River. Such habitat has been defined as places in the river where physical, chemical, and biological factors, including soil and water quality, sustain all life stages of fish and wildlife, including their reproduction (GLFC 1987; Environment Canada 1998).

Should we be concerned about habitat for fish and wildlife in the Detroit River? Yes, because the river also sustains us. Over five million of us live within an hour's drive of the river, drink water from it, and discharge our wastes into it. Many of us are employed at a business that depends on the river or enjoy its natural resources. However, we sometimes lose sight of our relationship with the river. Over time, our expanding population has reduced habitat of fish and wildlife, reduced their abundance, impacted water quality, and impaired our uses of the river and its natural resources (MDNR 1996).

Long before Europeans discovered it, the Detroit River was a source of food for Native Americans, such as the Wyandot of Anderdon Nation (see chapter 2). Long lots along the river ensured that each European settler had access to the river. In the 1870s, a commercial fishery employed thirty men, working day and night for months with seines, to harvest 45,000 spawning whitefish per year from shallow waters around Grassy Island (Milner 1874). People also ate waterfowl harvested from the river. In the 1800s, market hunters harvested ducks and geese from sneak boats with punt guns. Until recently, an annual muskrat festival fed many Gibraltar residents. Today, fish and wildlife populations in the river have dwindled to levels that preclude commercial harvest. Yet, thousands of people still fish and hunt on the river. One reason why fish and wildlife survive in the river is the large volume of clean water passing through it from Lake Huron. Another reason is the abundant wild celery that grows on the river bottom. Leaves of wild celery provide cover for young fish, and its starchy tubers are the preferred food of diving ducks that stop here as they migrate from Canada (Manny, Edsall, and Jaworski 1988).

We also use the river for recreation. Over 870,000 boats are registered in Michigan, about half of which are used on Lake St. Clair and the Detroit River. Many people fish for the estimated ten million walleye that ascend the Detroit River each spring from Lake Erie to spawn. Over 7,500 local members of the National Audubon Society in southeast Michigan enjoy bird-watching along the river.

Native Americans first navigated the Detroit River in canoes (Hinsdale 1927). In the 1700s, fur traders navigated the river with sailing schooners (Larson 1981). By 1820, cargo and passengers were carried on the river by a 135–foot paddle wheeler *Walk-In-The-Water* (Larson 1981). Around 1918, limestone rock outcroppings in the lower Detroit River were removed to facilitate shipping, and large runs of lake herring and lake whitefish that entered the river to spawn there ceased (Manny, Edsall, and Jaworski 1988). Thereafter, shipping channels were deepened throughout the river for commercial navigation. Maintenance dredging removes over a million cubic yards of sediment from these channels each year (Manny, Edsall, and Jaworski 1988). Waste discharges have contaminated much of the river bottom, damaging growth and reproduction by fish (Manny and Kenaga 1991). Equally damaging to wildlife and water quality have been discharges of oil (see chapter 6), nutrients (see chapter 7), and toxic substances (see chapter 8).

Development of the shoreline has been another important use of the river. It began in the 1790s with long lots that provided European settlers access to the river as they established their farms. In 1812, the Erie Canal opened and people flocked to the Detroit area. By 1827, industries and agriculture had modified the river shoreline to produce food, receive raw materials, and ship finished products to market (Dunnigan 2001). The usual practice was to build a bulkhead out from shore to water deep enough for cargo vessels. Then the area from shore over the marshlands to the bulkhead was filled with any available solid material. Gradually, nearly all the earthen, marshy shoreline along the Detroit River was replaced with bulkheads of steel, concrete, or wood that provided little or no habitat for fish and wildlife. Arrival of the railroad in the 1850s increased the flow of settlers and established salt and steel industries along the waterfront. By 1854, the Eureka Iron Works was in peak production of iron from ore shipped by water from Michigan's Upper Peninsula. Residential development for workers in these industries filled in the shoreline, creating the cities of Detroit, Ecorse, Wyandotte, Trenton, and Gibraltar. Large-scale concrete production for roads and structures accelerated urban growth. In the 1890s, more ships were built in Detroit than in any other American city. In 1897, a power plant supplied light, heat, and power to industry, businesses, and homes. Docks and storage facilities were built along the waterfront for shipping grain from local farms. Agriculture peaked in Wayne County in the 1890s at about 550,000 acres in production. Since 1910, agriculture has decreased as riparian land has been converted to urban and marina development. Now, only 3 percent of the original coastal wetlands exist along the Michigan shore, mostly on offshore islands (Manny 1998). In Canada, 13.5 percent of the shoreline is still agricultural (GDAHRI 1997).

Twenty-one islands in the river provide a serene environment, isolated from the hectic pace of urban life by the steady flow of the river. As Detroit enters the twenty-first century, the clear, blue water of the Detroit River sets the stage for international trade by high-tech industries in a global economy. Yet for many people, fish and wildlife in the river still serve as a measure of their quality of life.

How unique is habitat in the Detroit River? The river and western Lake Erie once contained one of North America's most diverse collection of freshwater clams (Goodrich and van der Schalie 1932). Wayne County still contains most of the remnant lakeplain prairies in the Great Lakes Basin (MNFI 1995). The 460–acre Wyandotte National Wildlife Refuge, the Canard River Marsh Complex in Ontario, and the river's islands are recognized globally for the biological diversity of their natural resources (Vigmostad 1999). For these and other biological reasons, the Detroit River was recognized as having significant biological diversity that should be conserved and enhanced by Canada and the United States under the 1992 United Nations Convention on Biological Diversity and by Michigan under its Biodiversity Conservation Act (Nature Conservancy 1994).

Ecologically, the Detroit River links the cold, sparsely populated upper Great Lakes region around Lakes Superior, Michigan, and Huron to the warmer, more densely populated lower Great Lakes region around Lakes Erie and Ontario. It also links St. Clair Flats, the largest freshwater delta in the Great Lakes, to Lake Erie, the most productive of the Great Lakes. The river connects residents of the area with thousands of miles of high-quality Great Lakes water. It contains vast fish spawning grounds (Goodyear et al. 1982) and supports one of our nation's most productive sport fisheries for walleye, bass, and muskellunge.

Scientists have demonstrated that the Detroit River is at the intersection of two major flyways for fish, birds, and insects, including:

- 117 species of fish (Manny, Edsall, and Jaworski 1988)
- 27 species of waterfowl that frequent Michigan's coastal wetlands (Prince, Padding, and Khapton 1992)
- 17 species of raptors, including eagles, hawks, and falcons (see chapter 11)
- 48 species of non-raptors, including loons, warblers, neotropical songbirds, cranes, and shore birds
- 35 species of dragonflies and butterflies (Holiday Beach Migration Observatory 1997a, 1997b, 1998)

In the North American Waterfowl Management Plan, the Detroit River was identified as "significant, international, waterfowl habitat of major concern" (USFWS/CWS 1986). The river was also nominated as a focus area for habitat restoration by the U.S. Fish and Wildlife Service and as a Biodiversity Investment Area by the U.S. Environmental Protection Agency and Environment Canada (Rodriguez and Reid 2001). It was also recognized as an important staging area for ducks by the North American Migratory Bird Conservation Commission. Partly for these reasons, President Clinton designated the Detroit River as one of the nation's fourteen American Heritage Rivers (see chapter 14).

The loss of coastal wetlands and other habitat in the Detroit River has been rapid. First explorers like Father Hennepin and Antoine Cadillac described the Detroit River as a pristine "paradise" with abundant edible fruits, lush meadows, forests, fish, and wildlife (see chapter 4). In 1815, the river shoreline consisted of a coastal wetland up to a mile wide along both sides of the river (figure 10.1). Vegetation types included submersed marsh, emergent marsh, wet meadow and shrub swamp, swamp forest and lakeplain prairie. Since 1815, the Detroit River ecosystem has undergone dramatic changes. Habitats for fish and wildlife are now degraded by contaminants, largely destroyed by shoreline and channel modifications, and greatly reduced in abundance and quality from historic levels.

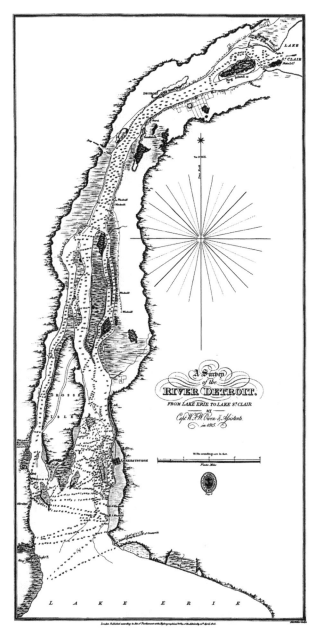

Figure 10.1. An 1815 map of the Detroit River showing coastal wetlands up to a mile wide along both sides of the river for most of its length, prior to shoreline development (Association of Canadian Map Libraries, Facsimile No. 20).

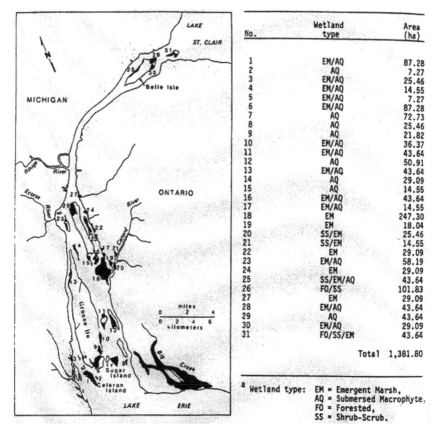

No.	Wetland type[a]	Area (ha)
1	EM/AQ	87.28
2	AQ	7.27
3	EM/AQ	25.46
4	EM/AQ	14.55
5	EM/AQ	7.27
6	EM/AQ	87.28
7	AQ	72.73
8	AQ	25.46
9	AQ	21.82
10	EM/AQ	36.37
11	EM/AQ	43.64
12	AQ	50.91
13	EM/AQ	43.64
14	AQ	29.09
15	AQ	14.55
16	EM/AQ	43.64
17	EM/AQ	14.55
18	EM	247.30
19	EM	18.04
20	SS/EM	25.46
21	SS/EM	14.55
22	EM	29.09
23	EM/AQ	58.19
24	EM	29.09
25	SS/EM/AQ	43.64
26	FO/SS	101.83
27	EM	29.09
28	EM/AQ	43.64
29	AQ	43.64
30	EM/AQ	29.09
31	FO/SS/EM	43.64

Total 1,381.80

[a] Wetland type: EM = Emergent Marsh,
AQ = Submersed Macrophyte,
FO = Forested,
SS = Shrub-Scrub.

Figure 10.2. Distribution of wetlands and large submersed macrophyte beds in the Detroit River, July 1982 (Manny, Edsall, and Jaworski 1988).

The largest habitat change has been encroachment into the river and hardening of the shoreline by the addition of sheet steel, cement walls, and fill material. Analysis of figure 10.1 reveals that 10.7 square miles (2,771 hectares) of coastal wetlands were present on the Michigan mainland in 1815. Analysis of 1982 landsat photographs (figure 10.2) reveals that only a tenth of a square mile (25.9 hectares) of coastal wetlands remained on the Michigan mainland by that year, mostly in the vicinity of Humbug Marsh. By 1982, more than 99 percent of the coastal wetlands present in 1815 along the Michigan mainland were converted to other land uses. Other losses of habitat included: removal of limestone spawning grounds for whitefish to create navigation channels; clearing of wooded areas for agriculture; and contamination of the water and bottom sediments by waste effluents. In the process, people lost benefits provided by wetlands along the river, such as flood control, protection from shoreline erosion, and removal of nutrients and sediment.

Progress in Addressing Environmental Problems

In 1985, the Detroit River was designated as an Area of Concern (AOC) by the International Joint Commission (IJC) to ensure that a Remedial Action Plan (RAP) was written to restore beneficial water uses (IJC 1985; Hartig and Thomas 1988; Hartig and Zarull 1992). In 1987, the protocol for the binational Great Lakes Water Quality Agreement was amended to define ecosystem integrity in terms of water-use impairments (Hartig et al. 1997). In 1991, the Detroit River RAP found that remaining fish and wildlife habitat did not meet fish and wildlife management goals (MDNR 1991). Loss of habitat was attributed to disruption of the physical, chemical, or biological integrity of the boundary waters, including wetlands (IJC 1991; table 10.1). In 1996, a RAP report defined objectives for restoring fish and wildlife habitat in the river (MDNR 1996). That report identified twenty candidate sites for habitat restoration in Ontario and Michigan and assigned high priority to conducting an inventory of remaining habitat for fish and wildlife in Michigan waters.

Assessment and Evaluation of Aquatic Habitats in Michigan Waters

In 1999, the U.S. Environmental Protection Agency funded the U.S. Geological Survey's Great Lakes Science Center to conduct an inventory of fish and wildlife habitat in Michigan waters. The project was titled "Detroit River Candidate Sites for Habitat Protection and Remediation." Scientists examined color aerial photography of the river and adjacent lands provided by the U.S. Army Corps of Engineers and noted the location of each green, undeveloped site on the mainland and all islands. They visited each site to:

- determine its latitude and longitude
- photograph its river shoreline and biological resources
- describe its landforms, vegetation, and wildlife
- assess its level of threat from development
- note uses of adjacent lands
- rate existing habitat on a scale from "pristine" to "highly degraded"
- classify the current uses and ecological condition of the land

TABLE 10.1

Comparison of Stage 1 RAP, IJC Stage 1 Review Comments, 1996 RAP Report, and Detroit River Canadian Cleanup Committee (DRCCC) assessment regarding beneficial use impairments

Use Impairment	Stage 1 RAP Conclusion	IJC Review Comments Stage 1 RAP	1996 RAP Conclusion	DRCCC Assessment (1998)
Restrictions of fish and wildlife consumption	Impaired due to PCB and mercury levels in certain fish	No other advisories exist; need hazard assessment for other contaminants in fish and wildlife	Impaired for fish	Impaired for fish
Tainting of fish and wildlife flavor	Not impaired	Need study to verify conclusion	Impaired for fish	Impaired for fish
Degraded fish and wildlife populations	Not impaired	Available data do not support conclusion; need definitive study	Not impaired for fish; unknown for wildlife	Impaired
Fish tumors or other deformities	Impaired	Concur	Impaired	Impaired
Bird or animal deformities or reproductive problems	Not impaired	Need study to support conclusion	Unknown	Impaired
Degradation of benthos	Impaired	RAP does not acknowledge link between sediment toxicity and degraded benthos	Impaired	Impaired

Restrictions on dredging activities	Impaired	Sediment toxicity results confirm this use impairment	Impaired	Impaired
Eutrophication or undesirable algae	Not impaired	No conclusive study available; need reassessment of available data to reach definitive conclusion	Not impaired	Not impaired
Restrictions on drinking water consumption or taste or odor problems	Impaired	Many locations within AOC could not be sites for drinking water intakes; existing sites have been threatened by spills	Impaired (taste and odor)	Impaired (taste and odor)
Beach closings	Impaired	Concur	Impaired	Impaired
Degradation of aesthetics	Impaired	Concur	Impaired	Impaired
Added cost to agriculture or industry	Not impaired	Limited documentation available; need user survey to make determination	Not impaired	Not impaired
Degradation of phytoplankton or zooplankton populations	Not impaired	Bioassays in Trenton Channel suggest impairment; need definitive study	Not impaired	Not impaired
Loss of fish and wildlife habitat	Impaired	Sediment toxicity results and violations of water quality objectives suggest impairment due to contaminants	Impaired	Impaired
Exceedance of water quality standards or objectives	Not addressed	Not addressed	Not addressed	Impaired

Scientists also examined public records to determine the name, ownership (public or private), area, assessed value, present zoning, zoning in the master plan, feet of river frontage, shoreline treatment, fish and wildlife resources, wetland classification, habitat quality rating, remediation potential, remediation planned, remediation completed, and possible sources of funding for remediation of each site. The project will yield five products, linked by a geographic information system (GIS):

1. a color map showing the location, assigned number, and name of each site
2. a computer database (GIS spread sheet) summarizing all information gathered about each site
3. sources of all the information gathered (USGS 1998, 2000)
4. photographs of each site
5. a written report that ranks all sites in priority for protection or restoration

The inventory revealed 104 sites totaling 3,436 acres in Michigan. Thirty-nine sites made up of 1,599 acres (44 percent of total area) are in private ownership, 65 sites of 1,837 acres (56 percent) are in public ownership, and 895 acres (26 percent) are contaminated brownfields. Less than 10 percent of the 32–mile riverfront is undeveloped, earthen shoreline.

These candidate sites are a starting point for balanced, sustainable use of the river for recreation, aesthetic enjoyment, and economic development. Every jurisdiction along the Michigan riverfront has provided data to the project and received copies of the color map. The success of the project will be measured by how much habitat at the candidate sites is protected or restored during implementation of the Detroit River RAP and other riverfront projects.

Setting Priorities to Protect and Restore Habitat in the Detroit River

The process of protecting and restoring habitat for fish and wildlife in the Detroit River will require natural resource planning and management agencies to set priorities. They may choose to follow these steps. First, state and provincial management biologists could agree on their natural resource vision for the river, i.e., how many of each species of fish, wildlife, and plants they desire to sustain in the Detroit River in perpetuity for public use and enjoyment. Then other natural resource organizations (e.g., the Audubon Society, Michigan United Conservation Clubs, and the Nature Conservancy) could add other species of desired plants and animals to the list. When the list of desired species has been reviewed by the public and is judged by all to be complete, they could move to the second priority.

Second, natural resource professionals could agree on how much of each kind of habitat once present in the Detroit River is needed to sustain in perpetuity each of the desired fish, wildlife, and plant species listed in the natural resources vision. Some species are year-round residents of the river and need different habitats to sustain themselves during each stage of their life cycle. Other species, such as diving ducks, are migratory and need only one or two kinds of habitat (such as beds of wild celery and resting areas away from human disturbance) during their brief stay on the river each year. Estimates of how much of each kind of habitat is needed could be based on available published scientific information and field experience of professional biologists.

The third step could be to determine how much of each kind of needed habitat remains in the river. Candidate sites for habitat protection and restoration are known for Ontario waters (OMNR 1993; Taylor 1998) and are being determined for Michigan waters (see above). Restoration of Detroit River habitats will likely proceed concurrently on several fronts: protection of natural areas that are still productive of fish and wildlife; restoration of natural areas impaired by physical modification and chemical contamination; and creation of new, uncontaminated habitat. It could be futile to protect and restore habitat for production of fish and wildlife if that habitat is not permanently protected from future development and contamination by toxic substances. It could be worse to restore or create habitat that attracts exotic, nuisance species at the expense of native, desired species of plants and animals.

Protection and restoration of habitat could begin with the small percentage of undeveloped shoreline that remains today. As brownfields are redeveloped, it may be possible to create habitat for fish and wildlife and protect the shoreline from erosion using soft engineering (Caulk et al. 2000). The highest priority lands for protection could be large, undeveloped, uncontaminated areas that still produce fish and wildlife, such as Stony, Celeron, Peche, and Round Islands, the Canard River Marsh Complex, and Humbug Island and Marsh. The next highest priority could be smaller islands in the "Conservation Crescent" around the southern tip of Grosse Isle, remnant coastal wetlands, and bottomland hardwoods. Protection of high-quality, riparian habitat in private ownership could be encouraged by local land trusts working in concert with state and federal programs, such as the Partners in Flight program of the U.S. Fish and Wildlife Service. Essential components of any mechanism for identifying and protecting habitat in the river will include:

- identification and evaluation of natural areas using scientific criteria
- realistic land-use planning by local jurisdictions and landowners
- adequate funds to purchase conservation easements on private lands
- enforcement of restrictions on filling and degradation of protected habitats (Taylor 1998)

Habitat restoration on public and private lands could be encouraged. Habitat restoration may be more easily accomplished on public lands because permits can be readily obtained, authority for projects is provided in existing legislation, and funding is provided in existing government programs. Most important, restored habitat on public land will not be developed later because development rights on public lands are restricted by conservation easements and deed restrictions. Nearly half of the remaining shoreline open space in Michigan waters of the Detroit River is privately owned. Those owners could be encouraged to protect their property from development if it has habitat value or potential. Although Michigan has provided funds to purchase development rights and establish conservation easements on prime farmland, no state funds are available to purchase development rights on private land along the Detroit River. A market-driven transfer of development rights program could protect habitat in sending areas in exchange for facilitated development of brownfields in receiving areas. Incentives for protection of habitat were included in legislation that recently established the Detroit River International Wildlife Refuge (Dingell 2001; MAC 2001). Private landowners in Michigan may get further information about habitat conservation by contacting a local land conservancy, local elected officials, or natural resource agencies.

In Ontario, a grant incentive program for habitat conservation has been proposed to the Environment Canada's Great Lakes Sustainability Fund but not yet implemented (Child 2001). Gifts of land that qualify as ecologically sensitive, also known as "ecogifts," have also become a useful tool in conserving threatened habitats for fish and wildlife in Ontario because they offer attractive tax advantages (Environment Canada 2001).

Industry and corporations that own land along the Detroit River are restoring habitat on their property and could perhaps do more, if they were organized like member corporations of the Wildlife Habitat Council along the St. Clair River. There, the council's Waterways for Wildlife Project recognizes corporations that manage their lands for wildlife habitat (WHC 1995, 2000). The shoreline of public lands, such as Grassy Island in the Wyandotte National Wildlife Refuge, Crystal Bay Island, and the Sugar Island Cut Dike could be modified to create coastal wetlands of bulrushes, shrub swamp, wet prairie, mixed hardwood swamps, and beech-maple forests like those present along the river at the time of European settlement (MNFI 2000). In such areas gently sloping, earthen shorelines could provide a gradient in water depth that allows coastal vegetation to survive changes in water level (Mitsch and Gosselink 1993; Maynard and Wilcox 1996; MNFI 1997). Water levels vary up to five feet in the Detroit River (Quinn 1981).

Choices about what, when, where, why, and how to restore habitat are best based on credible scientific information. Ideally, we should know in advance exactly what kind of habitat is needed to produce the desired number of each desired species listed by management agencies in the Detroit River natural resources vision. Why restore habitat if desired fish and wildlife species do not use it, if habitat essential for part of the life cycle of a desired species is not restored, or if undesirable (non-native) fish or wildlife occupy the restored habitat? Each habitat restoration project could be designed to sustain a target species of desired plant or animal. Ideally, the kinds of habitat needed by the greatest diversity of desired species could be restored first. Enough of each habitat used by each life stage of a desirable plant or animal species needs to be available for that species to sustain itself in the river. Progress in reaching benchmarks, objectives, and goals for protected, clean, and productive habitats for fish and wildlife could be assessed after each project of habitat protection or restoration is completed.

Some habitats in the path of economic development along tributaries to the Detroit River have been so altered by dams and water pollution that life stages of fish and wildlife species dependent upon them are nearly gone. To the extent that dams on the tributaries are removed and water pollution abated, those species could recover. In all habitat restoration projects, design concepts must be in harmony with natural forces inherent in the river. In the Mississippi River, such forces were harnessed successfully to create and manage productive habitats for fish and wildlife (Schnick et al. 1982).

Plenty of guidance exists for:

- restoring aquatic plant and animal communities (Morgan, Collicutt, and Thompson 1995; Environment Canada 1996; Wilcox and Whillans 1999; Pashley et al. 2000)
- setting restoration targets and location priorities (Environment Canada 1998)
- designing successful habitat restoration projects (OMNR 1993; Kelso and Hartig 1995; Caulk et al. 2000)
- controlling shoreline erosion (Fuller 1997)
- conserving native shoreline vegetation (Henderson, Dindorf, and Rozumalski 1999)
- restoring and protecting wetlands in private ownership (Cwikiel 1998)

Examples of successful habitat protection in the Detroit River include:

- purchase of 101–acre Stony Island (Ginnebaugh 1998) and 50–acre Peche Island by the state of Michigan and the city of Windsor, respectively
- protection of Ruwe Marsh near the Canard River (Tulen 1998)
- donation of the 115–acre Hennepin Marsh in the Trenton Channel by BASF Corporation to the Grosse Ile Nature and Land Conservancy
- donation of Mud Island by National Steel Corporation to the U.S. Fish and Wildlife Service
- transfer of 40 riparian acres on Gibraltar Bay from the federal government to the Township of Grosse Ile for a Nature Education Center
- purchase of numerous open spaces on Grosse Ile by the citizens of Grosse Ile Township through a dedicated tax millage
- purchase of Calf Island by the Nature Conservancy

Because of their large area, high biodiversity, and high level of development threat, Ojibway Shores just south of Windsor and the Humbug Marsh area in Michigan could be a high priority for protection.

Canadian biologists have established target amounts of contiguous forest cover required by deep-nesting birds and riparian vegetation to restore the Detroit River watershed to a healthy, functional level (DRCCC 1999). Since 1993, numerous successful habitat restoration projects have been completed along the Ontario shore, including Goose Bay Park on the Windsor waterfront, Turkey Creek channel improvements, stabilization of the river shoreline, cleanup and reforestation of the Little River watershed, and enhancement of the Canard River marshes (DRCCC 1999). To date in Michigan waters, only one habitat restoration project has been completed: removal of contaminated sediments from Monguagon Creek by the Michigan Department of Environmental Quality, with funding from the U.S. Environmental Protection Agency. A "Spirit of Trenton" project to landscape and restore habitat at five road ends along the Trenton Channel (City of Trenton 1998) is in progress. Feasibility studies were done by the U.S. Army Corps of Engineers for habitat restoration at Belle Isle, Hennepin Marsh, and the Black Lagoon, and creation of a marsh as part of the Conner Creek Combined Sewer Overflow Project (NWF 1999). However, few habitats have been restored to full productivity, decontaminated, and protected in perpetuity in Michigan waters of the Detroit River. During the 1990s, wild celery recovered throughout the river, due to greater water clarity and better light penetration (Manny and Schloesser 1999). Other native vegetation that provides cover and food for resident and migratory birds could be included in restorations of riparian habitat (see chapter 11). Lastly, habitats for spawning fish and refuges for resting waterfowl could be established in offshore waters of the river.

Delisting Loss of Habitat as an Impaired Beneficial Water Use of the Detroit River

Considerable thought has been given as to how the Detroit River could be delisted as an AOC (Hartig and Mikol 1992). Delisting loss of fish and wildlife habitat as an impaired beneficial water use could be possible when we achieve the following guideline of the IJC (IJC 1991; see table 10.1)— "[w]hen the amount and quality of physical, chemical, and biological habitat required to meet fish and wildlife management goals have been achieved and protected" (5).

Achievement of that guideline depends on the understanding and expertise of professional fish and wildlife biologists in state and provincial natural resource management agencies. Their collective knowledge about the amount and quality of habitat needed to sustain desired species in the river may someday be embodied in a comprehensive, binational Detroit River habitat management plan. Such a plan could identify and list how many of each desired species of fish and wildlife they are mandated to provide in the river for public use and enjoyment, and how much of each kind of habitat is needed to produce and sustain that many of each species. The list of desired species could be enlarged to include other plants and animals that are desired in the river by other natural resource organizations. Lastly, professional biologists, in concert with the four parties to the RAP process, could set realistic, achievable benchmarks for the protection and restoration of those habitats that everyone agrees will permit loss of habitat to be delisted as an impaired water use. Ideally, the process of restoring needed habitat would define benchmarks of progress that could be monitored in quantitative terms, such as acres of productive and uncontaminated habitat, and miles of natural earthen shoreline, that have been protected in perpetuity (Hartig et al. 1997). For five years after achieving the kinds and amounts of habitat needed to delist, the protected, restored, and created habitats could be monitored to verify that all those habitats remain protected, productive of fish and wildlife, and uncontaminated. Then, loss of habitat could be delisted as a water use impairment in the Detroit River.

References

Caulk, A. D., J. E. Gannon, J. R. Shaw, and J. H. Hartig. 2000. *Best Management Practices for Soft Engineering of Shorelines*. Greater Detroit American Heritage River Initiative. http://www.tellusnews.com.

Child, M. 2001. *Biodiversity Conservation Strategy Implementation for the Detroit River Area of Concern*. Project Proposal to Environment Canada's Great Lakes Sustainability Fund. Essex Region Conservation Authority, Essex, Ont.

City of Trenton. 1998. *City of Trenton Linked Riverfront Parks Master Plan*. Trenton, MI: City of Trenton, Engineering Department.

Cwikiel, W. 1998. *Living with Michigan's Wetlands: A Landowner's Guide*. Conway, MI: Tip of the Mitt Watershed Council.

Detroit River Canadian Cleanup Committee (DRCCC). 1999. *Detroit River Update Report*. University of Windsor, Ontario: Great Lakes Institute for Environmental Research.

Dingell, J. D. 2001. Detroit River International Wildlife Refuge Establishment Act. U. S. House of Representatives, Bill number H.R. 1230. 4 pp. plus map.

Dunnigan, B. L. 2001. *Frontier Metropolis: Picturing Early Detroit, 1701–1838*. Detroit: Wayne State University Press.

Environment Canada. 1996. *Planting the Seed: A Guide to Establishing Aquatic Plants*. Downsview, Ont.: Environmental Conservation Branch.

———. 1998. *A Framework for Guiding Habitat Rehabilitation in Great Lakes Areas of Concern*. Downsview, Ont.: Environmental Conservation Branch.

———. 2001. *The Ontario Ecogifts Handbook*. Downsview, Ont.: Canadian Wildlife Service.

Fuller, D. R. 1997. *Understanding, Living with, and Controlling Shoreline Erosion: A*

Guidebook for Shoreline Property Owners. Conway, MI: Tip of the Mitt Watershed Council.

Ginnebaugh, M. 1998. "Conserving Critical Habitats in the Conservation Crescent: The Stony Island Story." In *Rehabilitating and Conserving Detroit River Habitats,* ed. L. A. Tulen, J. H. Hartig, D. M. Dolan, and J. J. H. Ciborowski, 13–14. University of Windsor, Ont.: Great Lakes Institute for Environmental Research.

Great Lakes Fishery Commission (GLFC). 1987. *Guidelines for Fish Habitat Management and Planning in the Great Lakes.* Special Publication 87-1. Ann Arbor, MI: Great Lakes Fishery Commission.

Greater Detroit American Heritage Rivers Initiative (GDAHRI). 1997. *American Heritage Rivers Nomination.* Detroit: Rivertown Business Association.

Goodrich, C., and H. van der Schalie. 1932. "The Naiad Species of the Great Lakes." *Occasional Papers of the Museum of Zoology, University of Michigan* 238:8–14.

Goodyear, C. S., T. A. Edsall, D. M. Ormsby Dempsey, G. E. Moss, and P. E. Polanski. 1982. *Atlas of the Spawning and Nursery Areas of Great Lakes Fishes.* Vol. 8, Detroit River (FWS/OB 82/52). Washington, DC: U.S. Fish and Wildlife Service.

Halsey, John R. 1999. *Retrieving Michigan's Buried Past: The Archaeology of the Great Lakes State.* Bulletin 64. Bloomfield Hills, MI: Cranbrook Institute of Science.

Hartig, J. H., and G. Mikol. 1992. "'How Clean Is Clean'? An Operational Definition for Degraded Areas in the Great Lakes." *Journal of Environmental Engineering and Management* 2:15–23.

Hartig, J. H., and R. L. Thomas. 1988. "Development of Plans to Restore Degraded Areas in the Great Lakes." *Environmental Management* 12.3:327–47.

Hartig, J. H., and M. A. Zarull. 1992. *Under Raps: Toward Grassroots Ecological Democracy in the Great Lakes Basin.* Ann Arbor: University of Michigan Press.

Hartig, J. H., M. A. Zarull, T. B. Reynoldson, G. Mikol, V. A. Harris, R. G. Randall, and V. C. Cairns. 1997. "Quantifying Targets for Rehabilitating Degraded Areas of the Great Lakes." *Environmental Management* 21.5:713–23.

Henderson, C. L., C. J. Dindorf, and F. J. Rozumalski 1999. *Landscaping for Wildlife and Water Quality.* St. Paul: Minnesota's Bookstore.

Hinsdale, W. B. 1927. *The Indians of Washtenaw County, Michigan.* Ann Arbor, MI: George Wahr.

Holiday Beach Migration Observatory. 1997a. *26th Hawkwatch Season, Almanac for November 1997.* Essex, Ont.: Essex Region Conservation Authority.

———. 1997b. *26th Hawkwatch Season, Almanac for October 1997.* Essex, Ont.: Essex Region Conservation Authority.

———. 1998. "ERCA Internet Success: Hawkwatchers Access Web Page." *Northwind* 13.1 (winter): 9–17.

International Joint Commission (IJC). 1985. *Report on Great Lakes Water Quality.* Windsor, Ont.: Great Lakes Water Quality Board.

———. 1991. "Commission Approves List/Delist Criteria for Great Lakes Areas of Concern." *Focus* 16.1 (March/April): 1–5.

Kelso, J. R. M., and J. H. Hartig. 1995. "Methods of Modifying Habitat to Benefit the Great Lakes Ecosystem." Occasional Paper No. 1. Canadian Institute for Scientific and Technical Information, Ottawa.

Larson, J. W. 1981. *Essayons: A History of the Detroit District U.S. Army Corps of Engineers.* Detroit: U.S. Army Corps of Engineers, Detroit District.

Manny, B. A. 1998. "Ecological Restoration of Grassy Island and the Wyandotte National Wildlife Refuge in the Detroit River." In *Rehabilitating and Conserving Detroit River Habitats,* ed. L. A. Tulen, J. H. Hartig, D. M. Dolan, and J. J. H. Ciborowski, 18–21. University of Windsor, Ontario: Great Lakes Institute for Environmental Research.

Manny, B. A., and D. Kenaga. 1991. "The Detroit River: Effects of Contaminants and Human Activities on Aquatic Plants and Animals and Their Habitats." *Hydrobiologia* 219:269–79.

Manny, B. A., and D. W. Schloesser. 1999. "Recovery of Wild Celery *(Vallisneria americana)* in the Detroit River from 1985 to 1996." Paper presented to International Association of Great Lakes Research, Case Western Reserve University, Cleveland.

Manny, B. A., T. A. Edsall, and E. Jaworski. 1988. *The Detroit River, Michigan: An Ecological Profile.* U.S. Fish and Wildlife Service, Biological Report 85 (7.17).

Maynard, L., and D. Wilcox. 1996. "Coastal Wetlands of the Great Lakes." Background paper, State of the Lakes Ecosystem Conference 1996. Environment Canada and U.S. Environmental Protection Agency, EPA 095-D-96-001c, Chicago.

Metropolitan Affairs Coalition (MAC). 2001. *A Conservation Vision for the Lower Detroit River Ecosystem.* Detroit: Metropolitan Affairs Coalition.

Michigan Department of Natural Resources (MDNR). 1991. *Remedial Action Plan for Detroit River Area of Concern.* Lansing, MI: Surface Water Quality Division.

———. 1996. *Detroit River Remedial Action Plan Report.* Lansing, MI: Surface Water Quality Division.

Michigan Natural Features Inventory (MNFI). 1995. *A Survey of Lakeplain Prairie in Michigan.* Report to Michigan Department of Natural Resources, Land and Water Management Division, Coastal Zone Management Program, Lansing.

———. 1997. *Great Lakes Coastal Wetlands: An Overview of Controlling Abiotic Factors, Regional Distribution, and Species Composition.* Lansing, MI.

———. 2000. "Land Use circa 1800 County Base—by County [Wayne County]." Digital database map, edition 1.1. Lansing, MI.

Milner, J. W. 1874. *Report on the Fisheries of the Great Lakes; The Results of Inquiries Prosecuted in 1871 and 1872.* U.S. Commission of Fisheries, Annual Report, Washington, DC.

Mitsch, W. J., and J. G. Gosselink. 1993. *Wetlands.* 2d ed. New York: Van Nostrand Reinhold.

Morgan, J. P., D. R. Collicutt, and J. D. Thompson. 1995. *Restoring Canada's Native Prairies: A Practical Manual.* Chatham: Rural Lambton Stewardship Network, Ontario Ministry of Natural Resources.

National Wildlife Federation (NWF). 1999. *Report to the Great Lakes Fishery Trust for Creation of Fish and Wildlife Habitat at Conners Creek in the Detroit River.* Ann Arbor, MI: National Wildlife Federation.

The Nature Conservancy. 1994. *The Conservation of Biological Diversity in the Great*

Lakes Ecosystem: Issues and Opportunities. Chicago: U.S. Environmental Protection Agency.

Ontario Ministry of Natural Resources (OMNR). 1993. *Survey of Candidate Sites on the St. Clair and Detroit Rivers for Potential Habitat Rehabilitation/Enhancement.* Chatham, Ont.: Chatham Area Office.

Pashley, D. N., C. J. Beardmore, J. A. Fitzgerald, R. P. Ford, W. C. Hunter, M. S. Morrison, and K. V. Rosenberg. 2000. *Partners in Flight: Conservation of the Land Birds of the United States.* The Plains, VA: American Bird Conservancy.

Prince, H. H., P. I. Padding, and R. W. Khapton. 1992. "Waterfowl Use of the Laurentian Great Lakes." *Journal of Great Lakes Research* 18.4:673–99.

Quinn, F. H. 1981. "Secular Changes in Annual and Seasonal Great Lakes Precipitation 1854–1979, and Their Implications for Great Lakes Water Resource Studies." *Water Resource Research* 17:1619–24.

Rodriguez, K. M., and R. A. Reid. 2001. "Biodiversity Investment Areas: Rating the Potential for Protecting and Restoring the Great Lakes Ecosystem." *Ecological Restoration* 19.3:135–44.

Schnick, R. A., J. M. Morton, J. C. Mochalski, and J. T. Beall. 1982. *Mitigation and enhancement techniques for the Upper Mississippi River System and other large river systems.* Resource Publication 149. Washington, DC: U.S. Fish and Wildlife Service.

Taylor, S. 1998. "Identification and Protection Mechanisms for Detroit River Habitats." In *Rehabilitating and Conserving Detroit River Habitats,* ed. L. A. Tulen, J. H. Hartig, D. M. Dolan, and J. J. H. Ciborowski, 10–12. University of Windsor, Ont.: Great Lakes Institute for Environmental Research.

Tulen, L. 1998. "Ruwe Marsh Restoration Project." In *Rehabilitating and Conserving Detroit River Habitats,* ed. L. A. Tulen, J. H. Hartig, D. M. Dolan, and J.J.H. Ciborowski, 15–17. University of Windsor, Ontario: Great Lakes Institute for Environmental Research.

U.S. Fish and Wildlife Service/Canadian Wildlife Service (USFWS/CWS). 1986. *North American Waterfowl Management Plan: A Strategy for Cooperation.* Washington, DC: U.S. Department of Interior, Fish and Wildlife Service.

U.S. Geological Survey (USGS). 1998. *Content Standard for National Biological Information Infrastructure (NBII) Metadata, Workbook.* Washington, DC: Biological Resources Division.

———. 2000. *Proceedings of National Biological Information Infrastructure/National Park Service Metadata Training Workshop March 21–22, 2000.* San Antonio. Reston, VA: NBII Program Coordinator Office, USGS.

Vigmostad, K. E., ed. 1999. *State of the Great Lakes Islands.* Proceedings for the 1996 U.S.–Canada Great Lakes Islands Workshop, Michigan Office of the Great Lakes, Lansing.

Wilcox, D. A., and T. H. Whillans. 1999. "Techniques for Restoration of Disturbed Coastal Wetlands of the Great Lakes." *Wetlands* 19.4:835–57.

Wildlife Habitat Council (WHC). 1995. *St. Clair River Waterways for Wildlife Project Plan: A Public-Private Partnership to Rehabilitate and Enhance Habitat.* Detroit: Wildlife Habitat Council, Great Lakes Regional Office.

———. 2000. *Waterways for Wildlife: The St. Clair River Basin Project.* Summer 2000 Newsletter. Detroit: Wildlife Habitat Council, Great Lakes Regional Office.

Biodiversity of the Detroit River and Environs: Past, Present, and Future Prospects

James N. Bull and Julie Craves

A bald eagle, head and tail glowing white in the sun, arcs around the island. Legs outstretched it scrapes the water with a mighty whoosh, arcing back upward, a large fish writhing in its talons. Nearby on a point jutting out into the river, an osprey watches intently from its perch high in a tree. Dense shrubbery lines the water's edge. What seems a homogenous sea of dark greens on closer inspection reveals a gray and black sentinel. It's a black-crowned night heron. As the eyes adjust, there's another and another and another. At least thirty of these birds with swooped back crests guard the entire shoreline. A canoe paddle softly dips into the water, barely breaking the silence. As the craft glides along, dragging on the floating and sub-mergent vegetation, blue-wing teal, ducks with white crescent moon faces, are everywhere ahead. The canoe glides into a small inlet, the water covered with a blanket of lime green, tiny dots of duckweed plants, and large trees bending over in obeisance. Just rounding a bend, there's a clamorous splash, thunderous whoosh, and loud cackling, as the huge gray form springs over the canoer's heads—a great blue heron. A little further, there's rustling in a small group of trees. Looking up, the crowns are full of these three-foot tall lanky birds, looking very out of place and seemingly defying the strength of the small twigs that support them. The canoe turns another bend, and con-tinues. Now the sleek, white common egrets dot the trees. Steering back out toward open water, a yellow legs sandpiper probes a sandspit, while a com-mon tern glides by and is quickly gone.

The Everglades? The Okefenokee Swamp? Alaska? No. Amazingly, it's the view of two people canoeing on the Detroit River around Humbug Island and Marsh in the summer of 2000. The canoe trip seems like a step back in time.

Considering all the species seen on just one canoe trip, it should come as no surprise that in 1998, the U.S.–Canada State of the Lakes Ecosystem Conference (SOLEC) identified the Detroit River–Lake St. Clair ecosystem as one of twenty Biodiversity Investment Areas in the entire Great Lakes Basin with exceptional diversity of plants, fish, birds, and the requisite habitats to support them (Reid, Rodriguez, and Mysz 1999). Biodiversity Investment Areas are unique areas around the Great Lakes with exceptionally high ecological values that warrant special attention to protect them from degradation. Examples of these ecological values include:

- unique communities of plants and animals found in places such as coastal wetlands, dunes, and alvars (i.e., unique grassland, savanna, and sparsely vegetated rock barren ecosystems that develop on flat limestone or dolostone bedrock where soils are very shallow)
- concentrations of species of special interest, including rare, threatened, or endangered species
- exceptional species and habitat diversity
- high levels of ecological connectivity, both along the shoreline and to inland or offshore natural features

The Michigan Department of Natural Resources recognizes the Detroit River as having one of the highest diversities of fish and wildlife in all of Michigan. For example, as many as 29 species of waterfowl and 65 kinds of fish make their home along the river (Manny, Edsall, and Jaworski 1988). It is also a major migration corridor for fish, butterflies, waterfowl, raptors, and other birds. Indeed, the Detroit Audubon Society has observed over 300 species of birds in the Detroit-Windsor area; about 150 of them are known to breed here. These characteristics are part and parcel of the Detroit River's designation as an American Heritage River.

If the canoers beheld an amazing scene in the summer of 2000, what must this area have looked like when Detroit was founded three hundred years ago. Let's take a backward glance. In 1701 the pen of Detroit's (European) founder Antoine de la Mothe Cadillac described the river this way (Farmer 1890):

The Detroit is actually but a channel or river of medium breadth and twenty-five leagues in length according to my estimate. . . . Its borders are so many prairies [marshes no doubt] and freshness of the beautiful waters keep the banks always green. The prairies are bordered by long and broad rows of fruit trees that have never felt the careful hand of the vigilant gardener. Here also orchards young and old, soften and bend their branches, under the weight and quantity of their fruit, towards mother earth which produced them. It is in this land so fertile, that the ambitious vine [grape], which has never wept under the knife of a vine-dresser, builds a thick roof with its large leaves and heavy clusters, weighing down the top of the tree that receives it, and often stifling it with its embrace.

Under these broad walks one sees assembled by hundreds the timid deer and faun, also the squirrel bounding in his eagerness to collect the apples and plums with which the earth is covered. Here the cautious turkey calls and conducts her numerous brood to gather the grapes and here also their mates come to fill their gluttonous crops. Golden pheasants, the quail, the partridge, woodcock, and numerous doves swarm in the woods and over the country which is dotted and broken with thickets and high forests of full grown trees, forming a charming perspective. . . . The hand of the pitiless reaper has never mown the luxuriant grass upon which fatten woolly buffaloes of magnificent size and proportion.

There are ten species of forest trees, among them are the walnut, white oak, red oak, the ash, the pine, white-wood, and cottonwood; straight as arrows, without knots, and of prodigious size. Here the courageous eagle looks fixedly at the sun, with sufficient at his feet to satisfy his boldly armed claws. The fish are nourished and bathed by living water of crystal clearness, and their great abundance renders them nonetheless delicious. Swans are so numerous that one would take for lilies the reeds in which they are crowded together. The gabbling goose, the duck, the widgeon and the bustard are so abundant that to give an idea of their numbers I must the use the expression of a savage whom I asked before arriving if there was much game. "So much," he said, "that they draw up in lines to let the boats pass through." (Farmer 1890, 11)

Although today's river and the river of the past has and has had its spectaculars and its glories, the river's historical picture has not always been pretty.

It's 1961. A seven-year-old boy walks along an earthen dike jutting a half-mile out into the river from its origin near Belle Isle's water intake. He watches his footsteps as if life hung in the balance with every one. It did. Toe first, he gingerly puts his foot down. He looks down and around his shoe at the sun-dappled rocks, sand, and sparse vegetation, looking for a spot for his next toe. He is here with his father and Cranbrook naturalist Walter P. Nickell to band common terns. The men wear hard hats. About 1,000 nests, each with spotted eggs and young, blend well with the sand and stones, hence his caution. Hundreds of screaming terns, circling overhead and incensed at the intrusion, bomb the men below until clothes are painted all over with splotches of white. The other end of the terns is more dangerous. One after another they swoop down, jabbing the helmets with long sharp beaks, sounding like a percussion ensemble. On the boy, hard hat too big for him, they draw blood. Unfazed, the boy continues finding more nestlings to band. Sometime later one of these birds would be found dead in Chile, the band on its leg confirming its origins on the Detroit River.

The picture was quite different in late spring 1964. Wayne State ornithology student Gordon Peterson and Cranbrook naturalist Walter P. Nickell discovered that power mowers had run over hundreds of tern nests, chopping up hundreds of eggs and chicks into little pieces. Of the 360 nests they found, only 51 escaped the carnage. Gerald Remus, general manager of the Water Utility Board and long an outspoken enemy of the terns for "fouling the water intake" was suspected, but he denied responsibility. Terns were a hot topic at the next Detroit common council meeting where Nickell and Peterson testified in protest. Even Congressman Charles Diggs weighed in on behalf of the federally protected birds. At first the council seemed to favor paving the area to discourage further nesting by terns. In the end, the area was not paved, but still the terns never returned (Dowdy 1964; "Birds" 1964; Muller 1964). Nickell and crew continued banding terns on Bob-Lo Island and later on Mud Island until the mid-1970s.

Today the Belle Isle dike is lined with tall shrubs and the top of it is covered with short mown grass. Canada geese nest in this grassy area, most likely adding a much greater quantity of droppings to the water near the water intake than the terns did (Campbell 2000). Bob-Lo Island is a private community and Mud Island has become a small woodland. The terns, which are no longer "common," rarely nest on the Detroit River, and then only in small numbers. In fact, the common tern not only is considered a threatened species in Michigan, but also has been proposed for listing as endangered (Scharf 1991).

The common tern had most likely been nesting on islands and along the shores of the Detroit River for hundreds if not thousand of years. It hangs on, barely, but if the biota of the river environs were a mural of colorful tiles, one color, the terns, would be under-represented today. The mural that is the Detroit River, while in some areas still spectacular, has changed dramatically over the last three hundred years.

"Biodiversity" is a relatively new term coined in 1988 by E. O. Wilson, a pioneer in the new field of conservation biology (Jeremy et al. 1995). It refers to the richness and variety of the colorful tiles in the mosaic that is the earth, or any smaller portion of it—like the Detroit River area—and species are the different colors of tiles. Species are traditionally defined as groups of organisms that can reproduce and whose offspring can produce viable offspring as well. Once recognized only by external differences in features, species are often as recognized today by their DNA fingerprint as by wing shape or bud scar pattern. Through DNA analysis we have found organisms that are related to each other in ways we never dreamed. Conservation biology sprung up as a field to study this amazing variety of life forms and the ecosystems they depend on not long after the term "biodiversity" gained currency and concern was mounting about its loss. Scientists used to doing research became increasingly concerned that the earth's biodiversity was declining at a faster rate than ever before. Organized as a response to this complex crisis, this new field combines the rigorous science of the biologist and ecologist and insights about human behavior from the social scientist. Further, it brings together scientist and citizen in taking actions to halt the decreasing trend and reverse it if possible.

Biodiversity, often measured in terms of species richness (the number of species supported by an environment), is one critical measure. These species make up sub-pictures or ecosystems that are varied themselves: rainforests, freshwater and salt marshes, temperate forests and prairies, and the list goes on. So one can speak of the biodiversity not only of species but also of ecosystems. Because exotic or introduced organisms were not part of the original ecosystem mosaic of an area, native species richness is a more salient measure than species richness alone. But there is another dimension to the picture, alluded to above, that is equally important—species abundance. Just how many tiles of each color do you have? If you have too few, the color is hard to see in the mix. If too many, the color may be overwhelming.

Why is the planet's mosaic losing its colors? The threats to biodiversity are many: air and water pollution, introduction of exotic species, direct destruction (as with the terns), soil compaction, chemical additions to ecosystems, and global and microclimate change, but by far the most important factor has been loss and fragmentation of habitat. Causes for localized loss of biodiversity are much the same. The tern example is a good illustration. While mowing chicks down was direct destruction, it really was habitat loss that sealed their fate.

In order to understand the mosaic we see today in the Detroit River and its environs, a step back from the whole to look at some of the pieces of this priceless, dynamic mural will be helpful.

Habitats and Plants

While there has been no comprehensive study of presettlement vegetation, examining historic descriptions of the area, including the glowing report of Cadillac, gives us at least a general picture. By noting the ecosystems described in those accounts and comparing them to the ecosystems of the present, the fate of their constituent species can be inferred. Thus, our examination of plants will be framed by a look at four Detroit River area habitats over time: prairie, forest, wetland, and aquatic.

Prairies

Cadillac spoke of "the luxuriant grass upon which fatten woolly buffaloes of magnificent size and proportion," while other writings tell of bison so numerous that settlers considered making garments from their wool (Farmer 1890, 11). Another early account reports that "two leagues from Fort Detroit is an island called *Isle aux Dindes*. It is so called because turkeys are always to be found there. It contains but little timber, only prairie" (Farmer 1890, 7). Thus, both mainland and at least some islands had extensive prairies. They must have been extensive on the mainland to support the huge biomass of the bison herd. Most prairies have long since disappeared and along with them the bison, turkeys, and one of the richest ecosystems on earth. Without prairies, constituent plants like big blue stem grass, Indian switchgrass, prairie-fringed orchis, rattlesnake master, compass plant, and purple coneflower (to name just a few) have largely disappeared as well. Only in Ontario's Ojibway prairie and southeast Michigan's Sibley Prairie do pieces of this vast original habitat remain. Bison will never return even with a large-scale restoration or re-creation effort because they require such large expanses that are just not possible in what Schwartz (1997) calls "chronically fragmented urban habitats" (383, 392). The good news is that a few prairie plants have been found on Belle Isle and there are potential areas for prairie habitat on Detroit's east side. Prairie restoration is fairly well understood and the success stories in the Midwest are numerous. Remnant prairie plants often hang on along railroad tracks and occasionally in forgotten corners of open space. Some may still be holding on in Detroit waiting for favorable conditions to return.

Forests

Cadillac reported that the area has ten trees species. Among those he mentioned are red oak, ash, pine, walnut, white oak, white-wood, and cottonwood, "straight as arrows, without knots and of prodigious size." Other early accounts add maple, hickory, birch, beech, elm, butternut, cedar, basswood, and several conifers to the list of trees. The presence of bear, moose, lynx, and wolves indicates a forest wilderness of large proportions must have bordered the prairie swaths. Sugar, Hickory, and Bois Blanc (white wood, now Bob-Lo) Islands were named for their dominant trees: sugar maple, shagbark hickory, and either paper birch or aspen in the case of Bois Blanc. In 1813, when Tecumseh gathered his warriors on Bois Blanc, the white trees still stood, but when the U.S. patriots took charge in 1838, "the island was denuded of trees to get better range for . . . cannons" (Farmer 1890, 8). The orchards described by Cadillac and others were most likely planted by French voyageurs so that they might have access to these fruits later on during their travels.

Today, much of the original forest is gone. Pockets exist here and there notably on Peche Island, Belle Isle, Mud Island, Grosse Isle, the University of Michigan-Dearborn Natural Area, the Humbug Complex (Humbug Island and mainland), and parks surrounding the Detroit area. Some of the trees lining city streets are native swamp or riparian species that tolerate the harsh conditions there as well as they do in wet areas: silver maple, red maple, cottonwood, willow, and sycamore. Many others are exotics, which when planted near forests escape and compete with native trees. Tree of heaven from China and black locust from the Appalachians are foremost among them. Oak-hickory forests occupied the drier, well-drained sites, while beech-maple the more mesic areas, with many other ecosystem associations interspersed among them. Paw Paw and a few other southern species are at their northern extent in this region and still hang on. Peche Island, once a provincial park and now managed by the city of Windsor, boasts Carolinean forests similar to Point Pelee: a remnant southern forest with its characteristic Kentucky coffee tree and hackberry in northern refuge. The forests we have are wonderful, although a few might be in need of some exotic elimination or restoration of original processes. For instance, the periodic fires that once maintained the oak savannas have almost disappeared. Of the 1.8 million acres of forests in southeast Michigan during presettlement times, only 380,000 remain. We have lost 78 percent of southeast

Michigan forests, and a much greater percentage has been lost in the Detroit River watershed (Southeast Michigan Council of Governments 1999). The rest is being lost at an astounding rate. Suburban sprawl, as the Detroit area spreads out further and further north, east, and west, eats up more and more treed landscape making further declines in biodiversity a virtual certainty.

Wetlands

The prairies lining the Detroit River shoreline, of which Cadillac wrote, were most likely marshes, since "prairie" was used historically as a much more general term to denote areas dominated by herbaceous vegetation as opposed to trees and other woody plants. One of the first settlers of Wyandotte, George Clark, described that area this way: "Game of almost all descriptions was very plentiful in the rivers and marshes. The creeks and swamps teemed with fish, snakes, frogs etc." (R. P. Smith 1997). In 1812, the picture had not changed much according to Smith's (1997) account: "There were low marshlands along the whole expanse of the river and many of the creeks flowing into it were filled with high growing rushes. This along with the wooded areas just back from the shoreline and scattered corn fields gave protection to soldiers so that ambush was not uncommon."

Wetlands stretched not only along the shore of the Detroit River but as far inland as what is now Adams Avenue in Detroit, and perhaps beyond. When a piece of land just south of Adams was divided into city lots many protested, "Believe us Governor, no town will ever exist in these marshes" (Farmer 1890, 25). Grand Circus Park at the corner of Woodward and Adams was itself a marshland prior to becoming a city park. "As we rest in the shade of trees, enjoy the plash of the fountains, or watch children at play upon the lawns," Farmer wrote, "it is hard to realize that up to 1844 these parks were ponds and marshes, enlivened only the music of the bullfrog, and used as places of deposit for every kind of refuse" (73). With this general attitude toward wetlands and the tremendous growth of Detroit and southeast Michigan, 97 percent of Detroit River wetlands have been drained or otherwise lost through human impact (Tulen et al. 1998). Elizabeth Park, once known as Slocum's Island, was surrounded by marsh and was dredged so the area would become a true island. Owner Giles Slocum wanted this as a condition in 1919 when he donated the island to Wayne County. Swan Island (now Belle Isle) was reportedly "so covered with swamps that it was unusable until it was drained" (Smith 1997) and was surrounded with marsh. In 1873 the Detroit River system boasted 7,274 ha of wetlands, but by 1973 there were only 2,022 ha left (Hartig 1993).

Starting at Windmill Point, the northern extent of the Detroit River and moving south, the shoreline is encased in steel seawalls and development almost up to the riverbank. It isn't until just below the Detroit Edison plant at Solutia's property in Trenton that a small pond and marsh are found that were once used for chemical disposal. Beyond that is the last remnant mainland coastal wetland on the river's U.S. shoreline, the 465–acre Humbug Complex including the 100–acre Humbug Marsh, one of the few depositories of biodiversity left along the river, and perhaps one of the richest (Tulen et al. 1998). The threatened fire pink occurs in the adjacent upland and wetland borders, while the showy pink blossoms of the swamp rose mallow, a state species of special concern, dots the marsh with color in late summer and early fall. With the large number of sedges, rushes, bulrushes, and pond weeds on the state threatened list, it is possible that a trained botanist might find a few of these species hanging on as well. Hennepin Marsh is the last vestige of the once extensive wetlands on Grosse Ile, the river's largest island.

Crossing the river and going back north, we come to Canard River Marsh. This is the crown jewel of the Canadian side—the largest and most biologically significant wetland on the Detroit River. In fact this tributary was named for the large concentrations of ducks that assemble in its estuary and adjacent marsh. Although much smaller, the wetlands surrounding Grass Island and the Detroit River Wetlands near Fighting island are rich pockets of biodiversity as well. The Humbug Complex and these Canadian areas are threatened with development that would certainly compromise their value as wildlife habitat, if any habitat remains at all.

Although standing trees differentiate swamps from marshes in modern ecology, historic accounts use the word "swamp" for almost any wet area. In presettlement times the seven-county southeast Michigan area had almost 800,000 acres of swamps, but by 1995 only about 42,000 remained. Ninety-five percent of southeast Michigan's forested wetlands have been lost, and the loss in the smaller Detroit River area would be much closer to 100 percent (Southeast Michigan Council of Governments 1999). Remaining swamps, even rarer than marshes, are found on Peche Island, Belle Isle, and some small patches on Humbug Island.

With only 3 percent of the original Detroit River wetlands left, many of the plant species they supported are either gone or their population size is a tiny fraction of what it once was. Draining was the major destructive force, but other forces are more insidious. Introduced carp digging in the mud, making the water turbulent, can severely degrade a marsh. The alien purple loosestrife, introduced as an ornamental from Eurasia, poses another major threat, replacing native marsh plants and forming dense stands of beautiful lavender flowers, that are of little use to wetland fauna. Humbug Marsh was one of the only marshes that had not been invaded by purple loosestrife, but that changed in the last two years after a developer cleared some of the land and wetland border and brought in soil from other locations.

Although native and probably present for centuries in North America (Lapin and Randall 1993), the common reed in recent years has become invasive and has replaced other wetland species. Because, like loosestrife, it is of little or no value to wildlife for food or shelter, its invasiveness has led to declines in species richness and abundance in many remaining wetlands. It is not fully understood why this plant has become so invasive, but it is thought that a non-native strain was introduced into the United States; global warming and disturbance of marsh soils may also be contributing factors (Lapin and Randall 1993; Long Point Bird Observatory 2000). One characteristic of the common reed, which gives it an advantage in disturbed areas, especially along roadsides, is its tolerance of salt (Gleason and Cronquist 1963). The impact of drainage of salt from roads has dramatically increased along with ever-expanding development.

Aquatic Environment

Early accounts did not fully describe the aquatic flora, but the reports of superlative fishing in the area indicate that it must have been diverse and robust. Twenty types of submergent macrophytes are known to occur in the Detroit River in beds made up of 2 to 11 species, and usually above 7 meters in depth. Where the depth is 3.7 meters or less, around 72 percent of the area is covered with submergent vegetation (Manny, Edsall, and Jaworski 1988).

There are 82 species of phytoplankton in the Detroit River, with diatoms and bluegreen algae the most abundant. The Detroit River is notable for having the third highest number of diatom species in the whole Great Lakes system. Wild celery, musk grass, species of narrow-leafed potamogeton, Eurasian watermilfoil, and water stargrass are the most common, in that order. In any given location in the river there may be seasonal succession of species, so that a different plant species is dominant at different parts of the growing season (Manny, Edsall, and Jaworski 1988).

Eurasian watermilfoil, endemic to Europe, was introduced accidentally. It forms thick underwater mats stretched over large areas, impeding recreation and crowding out native submergent macrophytes. Higher nutrient levels exacerbate its growth. One of the linchpins of its success is its ability to propagate from stem fragments; a single stem fragment can be founder of a large mat producing colony. Stem pieces stuck on boat propellers, hulls, various kinds of fishing and boating equipment, and even on clothes can readily spread this invader to new water bodies. Mechanical clearing can make the problem worse because of all the stem fragments it creates (University of Minnesota Sea Grant 2000). Since herbicides can eradicate native species along with the invader, a native milfoil weevil *(Euhrychiosis lecontei)* is being studied as a possible biological control. Its preference for the Eurasian watermilfoil over native milfoils makes it an especially good candidate. Weevil larvae attack and eat the merristem, then bore through the stem where they feed on the cortex, which usually causes the plant to sink. Its effectiveness has varied in different sites, and so far there is no basis for prediction of how well it will work in any given area.

Another exotic submersed aquatic macrophyte, curly-leaf pondweed, was first discovered in the Detroit River in 1951 (Manny, Edsall, and Jaworski 1998). Its impact is similar to that of Eurasian watermilfoil, and it has the same habit of reproducing vegetatively from small fragments (University of Minnesota Sea Grant 2000). *Nitellopsis obtusa* is the most recent exotic macrophyte to be found in the Detroit River. Its occurrence only in waters used by commercial vessels suggests that it arrived in the Great Lakes from Eurasia by that conveyance. To date it has been found only at Belle Isle and Point Hennepin and has not reached a high enough population level to cause problems (Manny, Edsall, and Jaworski 1988), but further impact on native macrophytes and aquatic ecosystems is probable.

Invertebrates

There is little mention of invertebrate species in historic accounts. However, because early accounts of Detroit agree on the large variety and wide extent of several different ecosystem types (prairie, marsh, etc.), the large variety of vertebrate species, and their extraordinary abundance, a rich presettlement invertebrate fauna can be inferred. Although information is lacking on most invertebrates groups in the Detroit River area, with regard to class mollusca (mussels, snails, clams, slugs, etc.), analysis of some recent data (1980–2000) yielded important findings.

The zebra and quagga are exotic mollusks that entered the Great Lakes system through ballast water discharged to the Detroit River in 1986 and

1991, respectively (Schloesser et al. 1998). Surveys from before and after these infestations demonstrate how quickly biodiversity and species abundance can change. Surveys of mussels on the river bottom conducted in 1982–83 and 1992–94 noted the number of live and dead native and exotic species. Live individuals of native mussel species (family Unionidae) made up 97 percent of the samples prior to the exotic mussel infestations, but only 10 percent afterward. At one station the decline went from 84 percent to just 3 percent live native unionids. Twenty percent of the drop in unionid species richness came after the second exotic mollusk, the quagga, entered the Detroit River, and it happened over only a two-year period (Schloesser et al. 1998). The drop in live native mussels species in the samples averaged 7.7 percent per year during the first ten years of the zebra mussel infestation, but the drop accelerated to an average of 10 percent per year when the quagga was added to the system. While the rate of drop with the two mussels is less than would be expected if both mussels had an equal impact, the second mussel appears to have accelerated the drop in species richness by an average of 2.3 percent per year. Clearly there was more damage to mollusk species richness in the Detroit River with two exotic mussels present than there had been with only one. At least ten native mussel species in the river (eight uncommon and two common) have been extirpated since the introduction of these two exotic mussels—a decline in species richness of 45 percent (Schloesser et al. 1998). Thus, in just twelve years the Detroit River has lost almost half of its native mollusk species.

Three other invertebrates deserve special attention: spiny waterflea, *Daphnia lumholtzi,* and rusty crayfish. The first two species are thought to have entered the Great Lakes through ballast water. Spiny waterflea, a zooplankton native to Europe, was first found in Lake Huron in 1984. Its spiny tail makes it difficult for fish to eat, using up a fish's much needed energy and impacting feeding. In addition to its impact on fish, native zooplankton decreased in numbers as they were consumed by the larger waterflea, while the waterflea's numbers skyrocketed. Are native zooplankton species the next to join the threatened species list? Unfortunately the waterflea travels well on fishing reels and fishing line so inland lakes and streams are in danger of infestation as well (U.S. Fish and Wildlife Service 2000). Africa, Asia, and Australia were the home of *Daphnia lumholtzi* before it entered the Great Lakes with aquarium fish from Africa. It impacts the ecosystem in very similar ways to the spiny waterflea. Rusty crayfish, native to Tennessee, Kentucky, and Ohio, most likely entered the Great Lakes from leftover bait thrown in the water to feed the fish that would not be caught that day. It reduces aquatic plant cover for fish and competes with native crayfish (U.S. Fish and Wildlife Service 2000). It is not clear whether the rusty crayfish has entered the Detroit River as yet. If not, it surely will not be long.

Fish

Cadillac and later George Clark commented on the great numbers of fish in the Detroit River and its many marshes, while other accounts indicate that lake whitefish, pickerel, pike, perch, black bass, catfish, sunfish, and bull-heads were abundant (Farmer 1890). Large spawning runs of lake whitefish, lake herring, and lake trout were well-known in the Detroit River. As a result of this substantial fishery and increasing numbers of immigrants, a major commercial fishery developed in the river during the 1800s and was sustained up until the early 1900s. Professional fishermen, using seines and lines, harvested numerous lake whitefish, lake herring, yellow pike, yellow perch, and lake trout. Whitefish was the most abundant and most prized, going for as little as $3.50 per barrel in 1818 and "boatloads sold for 50¢". Near Grosse Ile, 25,000–30,000 whitefish were caught in 1824. One report states that 15,000 whitefish were caught in five hauls of just one seine. Beginning in 1873 whitefish were stocked from state fish hatcheries to augment natural reproduction. Barrels were sent by the thousands to markets in New York. In September 1826, a fish hatchery was established in Detroit at Atwater and Dequindre, raising as many as 45 million fish, most released in the river. Large numbers of fish, "from the half-pound Perch to the one-hundred-twenty pound Sturgeon—(were) caught yearly" (Farmer 1890, 16).

The large fish spawning runs ceased in the late 1800s and early 1900s because rock outcroppings, used for spawning, were destroyed by construction of the shipping channel or spawning habitat became degraded by water pollution (H. M. Smith 1917; Manny, Edsall, and Jaworski 1988). The Detroit River was also an important spawning ground for lake sturgeon. Both Native Americans and European settlers benefited from the lake sturgeon fishery. However, the lake sturgeon population soon became depleted. Its decline was attributed to overfishing, degradation, and loss of spawning habitat (Manny, Edsall, and Jaworski 1988).

By 1952 lake whitefish, pickerel, and sturgeon had apparently disappeared from the Detroit River (at least in the vicinity of Stony and Sugar Islands) (Farmer 1890; Cooper 1952). Carp and goldfish, both exotics, had become part of the aquatic fauna by this date. Carp, introduced in western

Lake Erie in 1883, has wreaked havoc with native species and environments while becoming one of the most numerous and ironically commercially valuable fish. Carp destroy wild celery beds important for canvasback and redhead ducks (Manny, Edsall, and Jaworski 1988) and muddy the water as they dig, interfering with gill function, photosynthesis, and temperature.

There were 47 species of fish identified in a one-week fish survey in September 1952. Twenty-five years and 30 or more years later (surveys in 1977 and 1983–86) the number of species had not changed much, declining by only one to 46, but the ecosystem had changed dramatically (Manny, Edsall, and Jaworski 1988; Haas et al. 1985; Muth, Wolfert, and Bur 1986). Sixteen species found in 1952 were not found in the later surveys, and there were fifteen new species. Among them, alewife, rainbow smelt, and sea lamprey were unintentional introductions from saltwater environments, entering when the Welland Canal provided access to the upper lakes, previously prevented by Niagara Falls. Chinook and coho salmon, introduced intentionally for sportfishing and to help control alewife and smelt, filled the niche once occupied by whitefish.

One species native to the Great Lakes that was added to the list in 1977 is worthy of note—the pugnose minnow. It had not been recorded in 1952, but juveniles were found in a 1977 survey in the Gibraltar Bay area. It is currently considered a threatened species in Michigan (Michigan Department of Natural Resources 1998). Since this minnow prefers sluggish weedy waters (Hobbs and Lagler 1958), its decline may be related to the disappearance of coastal marshes, and its survival may be attributed to the quality of the marshes that remain at Gibraltar Bay and the Humbug Complex.

Of the 65 species of fish Manny, Edsall, and Jaworski (1988) list as possible for the Detroit River, 39 are known to spawn in it. Several others use spots in the river as nursery areas: Gibraltar Bay and the Humbug Complex foremost among them. One fish that had been extirpated in the Detroit River and on the decline is apparently on its way back since the late 1990s—lake sturgeon.

While the trends in fish biodiversity are complex, it is clear that native species have been lost and many that remain are less abundant. Loss of coastal marshes and fish aversion to the seawalls that replaced them contributed more to the disappearance of native fish species and declines in their abundance than pollution or any other factor (Manny, Edsall, and Jaworski 1988). Exotics that compete with them (e.g., alewife) or parasitize them (e.g., sea lamprey) would probably be a close second.

Mammals

In addition to large herds of bison, early accounts tell of numerous elk, moose, beaver, wolf, bear, rabbit, otter, lynx, bobcat, and muskrat. In 1830, dwellers in the growing municipality could often still hear wolves howling beyond the city limits. In fact, bounties on their hides accounted for use of the majority of the taxes collected at that time (Farmer 1890). As the forests were cleared, bear, moose, wolf, and bobcat retreated further north in Michigan, the lynx on into Canada.

Demand for beaver pelts drove the French occupation of the Great Lakes region (see chapter 4). Detroit was founded as a fur trading post where voyageurs from the upper lakes would bring their furs for later transport to markets. Beaver furs, in high demand for fashionable hats in Europe, brought the highest prices. "Trapper's Alley" in Detroit's Greektown dates from this era and takes its name from the central place it played in this main engine of Detroit's early economy. Beaver are no longer found in the Detroit River area, but they do occur along Lake Erie and the Cuyahoga River in Ohio and further north in Michigan. The beaver spontaneously reintroduced itself to the Cuyahoga Valley in the early 1980s and has now spread north to the Cleveland Flats area, where it can be seen making bank dens and swimming along the river in the heart of urban-industrial Cleveland. Pollution cleanup efforts and the preservation of the Cuyahoga Valley, including letting nature reclaim former developments, are possible causes. Riverbanks lined with shrubs and trees, some overhanging the river, are almost continuous even through the most industrial areas, which may be another factor in the beaver's return. Based on Cleveland's experience, there may yet be hope for some beaver to return to the Detroit area, this time without feeling the impact of the fur trade.

Of the species listed above, the well-adapted cottontail rabbit alone is doing well and may even be on the upsurge. The white-tailed deer is the one large animal that is flourishing in outlying rural areas and suburbs that currently have more woods and fields. Overpopulation in these few remaining wild or semi-wild areas and the consequent disappearance of native flora have caused the Huron Clinton Metroparks, like many urban parks around the country, to institute control measures, including limited hunting and sharp-shooting, both of which have been controversial. Exotic species, which have been around for almost as long as human habitation, include the house mouse and Norway rat. A more aesthetic exotic, which is popular with people but may be impacting native flora on Belle Isle, is the European fallow deer.

Birds

Passenger pigeons roosted in such large numbers around the area in Detroit's early days that "hundreds could easily be killed with a walking stick" (Farmer 1890, 11). As late as 1850 wild turkeys and quail were abundant at the edge of town and found their way into the city from time to time. Ducks and geese migrated in large numbers over the area. Eastern meadowlarks, American robins, brown thrashers, bob-o-links, and ruby-throated hummingbirds were common in the fields and woods just beyond the city's borders (Farmer 1890, 12).

Today the Detroit River is still known for its richness and abundance of birds. Both its geographic location and intrinsic habitat characteristics attract birds throughout the year. Despite degradation, the lower river is considered an essential enough congregating point for waterfowl to be designated as a globally significant Important Bird Area by Bird Life International and its Canadian partners, the Canadian Nature Federation and Bird Studies Canada (Bird Studies Canada 1999).

The Importance of the Detroit River to Migratory Birds

Southeastern Michigan is located at the convergence of the Mississippi and Atlantic Flyways, two of the four major bird migration routes in North America. The Detroit River, with its approximately north-south orientation, is an important corridor along this flyway.

Raptors

On September 17, 1999, hawk counters for Southeast Michigan Raptor Research, located at Lake Erie Metro Park in Wayne County, were prepared for the peak of the annual broad-winged hawk migration. The previous one-day total for broad-wings at this location, just south of the mouth of the Detroit River, was a seemingly unbeatable 399,372, set in 1994. Yet the final tally on this autumn day ended up a jaw-dropping 555,371 (Tessen 2000). Later in the season, Halloween Day brought an additional treat: 48 golden eagles, which helped boost the 1999 fall count for this primarily western species at Southeast Michigan Raptor Research to 245. These totals were record counts for golden eagles in eastern North America (J. Schultz, personal communication).

Clearly the Detroit River is a prime migration highway for migrating raptors, especially in the fall. Reluctant to cross large bodies of water, they follow the Great Lakes shorelines until reaching the narrow crossing at the Detroit River. Fifteen species are regularly counted each year at Southeast Michigan Raptor Research and Holiday Beach Migration Observatory, just southeast of the mouth of the Detroit River in Essex County, Ontario. Table 11.1 gives recent mean annual totals for four raptor species counted at these sites. Total numbers of raptors counted annually at these two sites combined averages between 200,000–300,000 birds, one of the largest concentrations in the eastern United States.

Waterfowl

Cadillac talked about the profusion of swans, "chattering goose," duck, and teal. For nearly all of recorded history, the Detroit River has been a traditional staging area for migratory waterfowl, both during migration and over winter. Shallow water and rapid currents (and, since the advent of the industrial era, warm effluents) keep much of the river open all year. These factors, plus abundant submergent vegetation, especially wild celery, attract thousands of ducks (figure 11.1). Wild celery is particularly important for canvasbacks, redheads, and both greater and lesser scaup (Wooley 1998). In March 1941, an estimated 400,000 ducks were counted along the east side of Grosse Ile (Miller 1943), and in the mid-1950s, fall and early winter canvasback counts in the Detroit River were typically more than 200,000 (Schloesser and Manny 1990).

TABLE 11.1
Mean Number of Selected Raptors Counted During Fall Migration at Two Locations on the Detroit River, 1990–99

	SMRR[1]	HBMO[2]	Both locations
Turkey Vulture (*Cathartes aura*)	21,730	16,338	38,068
Bald Eagle (*Haliaeetus leucocephalus*)	82	55	137
Broad-winged Hawk (*Buteo platypterus*)	193,376	51,421	244,797
Red-tailed Hawk (*Buteo jamaicensis*)	5,871	7,289	13,160[1]

Notes: [1]Southeast Michigan Raptor Research, Wayne County, Michigan (P. Cypher, personal comment)
[2]Holiday Beach Migration Observatory, Essex County, Ontario (D. Benoit and R. Petitt, personal comment)

Figure 11.1. The lower Detroit River is a major feeding, nesting, and resting ground for waterfowl, including lesser scaup (Michigan Department of Natural Resources).

At this point, the Detroit River's waterfowl prosperity began to falter. Wild celery beds in the lower river declined 72 percent between 1950 and 1985. Reasons for the decline are unclear, but pollution (from waste oil, industrial and municipal effluents) is a likely candidate (Schloesser and Manny 1990). Considering that the Detroit River had already experienced pollution prior to the original wild celery survey in 1950, the historical numbers of wild celery beds and the abundance of ducks feeding on them must have been truly staggering.

As the wild celery declined, it resulted in a concomitant decline in the number of canvasbacks using the river, with fewer than 2,900 counted by 1985 (Schloesser and Manny 1990). The average number of canvasbacks counted for the period 1990–99 is around 10,000, but numbers have varied from fewer than 200 to a maximum of 21,000.

The annual duck harvest in 1941 for the lower Detroit River was 44,500 (Miller 1943). Wayne County duck harvests held at about 11,000 annually for 1961–80 (Manny, Edsall, and Jaworski 1988). Totals for selected species of diving ducks counted on the Christmas bird counts from 1990–99 are shown in table 11.2. Since these are one-day, ground-based

TABLE 11.2

Mean Number of Selected Diving Ducks Counted on Two Christmas Bird Counts Covering Portions of the Detroit River, 1990–99

	Detroit River MI-ON		Rockwood MI[2]	
	Annual Mean		Annual Mean	
	total birds	per 75 PH[1]	total birds	per 75 PH
Canvasback	2822.0	2786.6	7733.9	6886.4
Redhead	457.7	382.1	371.6	286.1
Scaup sp.	80.0	82.4	454.4	345.1
Common Goldeneye	518.4	510.2	1000.4	832.5
Common Merganser	445.7	465.3	4139.0	3704.5

[1]Per 75 party-hours (PH), a standard measure of counting effort; average number of PH per year for these counts = ~75.
[2]Rockwood count circle also includes Lake Erie shoreline from Pointe Mouillee State Game Area through Fermi Powerplant.

counts not utilizing water or aerial searches, they certainly undercount waterfowl. Still, a doubling or even quadrupling of these figures does not approach totals noted in the 1950s, much less the numbers likely present historically. Loss of habitat and preferred forage are surely to blame.

Landbirds

The Rouge River Bird Observatory at the University of Michigan-Dearborn, located in a mosaic of forest and fields along the Rouge River that is tributary to the Detroit River, has been studying the importance of urban natural areas to birds, especially migrants, since 1992. Over 250 bird species have been recorded (Craves 1996; Rouge River Bird Observatory 2000).

Over 80 percent of these are migrants (Craves and Gelderloos 1996), primarily songbirds. The availability of appropriate stopover habitat for migrant birds is critical to successful migration, and migration conditions may play a major role in the health of migratory bird populations (Moore et al. 1995; Moore and Simons 1989; Yong and Moore 1997; Yong et al. 1998). Nearly half of Michigan's migrant species have used the area as a stopover site to rest and refuel; most stayed more than one day and most gained weight (Craves and Gelderloos 1996). This illustrates how crucial stopover sites can be in urban areas, especially in regions along major flyways.

Still, migrant birds are not as abundant today as they were even at the turn of the twentieth century. On May 2, 1905, an observer on Belle Isle noted: "I believe I never saw so many birds in spring before—it simply tired one. Warblers were present in great numbers everywhere" (Swales 1893–1915).

The following case illustrates this well. Although the blackburnian warbler is considered a fairly common migrant in Wayne County (Craves 1996), as it was in the early 1900s (Cook 1893; Swales 1893–1915), the numbers observed have plummeted. While Wood (1909) found as many as 260 on one outing along the Rouge River, the highest statewide total for this flame-throated bird at the peak of spring migration in the 1990s was 82.

The Importance of the Detroit River to Breeding Birds

Marsh Birds

Historically, many wetland-dependent bird species nested in the marshes that once lined the Detroit River. As most of the river's wetlands have declined, so have the number of breeding birds. A sobering example is the story of Grassy Island. Mallard, wood duck, American black duck, northern pintail, and redhead nested in this area along the lower Detroit River as late as 1968. By 1983 when Michigan's new breeding birds atlas was published, there were no records of nesting in this area by pintails or redheads on either side of the river, and no nesting of blue-winged teals or black ducks on the American side (Manny, Edsall, and Jaworski 1988; Brewer, McPeek, and Adams 1991).

Terns and Gulls

In addition to habitat loss and outright destruction, a serious cause of common tern colony abandonment is competition from gulls, particularly ring-billed gulls, which are opportunistic and adapt readily to human-altered habitats (Ludwig 1991). Ring-billed gulls were absent or scarce as nesting birds in the Great Lakes until at least 1926, but in 1960 they were found nesting in the Detroit River (Ludwig 1962). In 1976–77, there were 5,000 ring-billed gull nests on Mud Island, nearly 2,000 on Grassy Island, and 20,000 on Fighting Island by 1984 (Blokpoel and Tessier 1986).

Concurrently, there was a 67 percent decline in nesting common terns in Michigan between 1960 and 1985 (Scharf 1991). In 1995, 33 tern nests were found on Fighting Island (Bird Studies Canada 1999), but BASF Corporation's continued efforts at restoration on Fighting Island have resulted in an increase in ring-billed gulls by over 230 percent since 1991, with a current population estimate of 350,000 (Lanigan 1998). In this case, there has been no literal "net loss" of biodiversity in terms of species richness, while its abundance has declined precipitously. We have nearly lost a species with specific needs and gained a species considered by many to be a pest.

Landbirds

For landbirds, even the area as urbanized as that along the Detroit River has the potential for high avian biodiversity. Evidence of nesting for over 80 bird species was noted by the Rouge River Bird Observatory at their site; 55 of them nest regularly, excluding waterfowl (Craves 1996). Following a formula given by Galli, Leck, and Forman (1976) for predicting species richness in forest islands, this site should have 50 breeding species based on its area (Craves and Gelderloos 1996). Again, this is an example of the importance of natural areas to birds in the Detroit River region and the diversity found there.

Unfortunately, the richness and particularly abundance of birds in the area today is far less than what was recorded formerly. Since 1915, about 70 percent of the bird species identified as regularly nesting in Wayne County have declined or are no longer present as breeding species (Craves, forthcoming).

Excellent insight on the diversity and abundance of birds in the Detroit area is found in twelve unpublished journals kept by Detroit resident and early ornithologist Bradshaw Swales (Swales 1893–1915). Although much of his writing was published, his daily journal entries provide data that at the time were not noteworthy enough for publication.

Swales noted that on Belle Isle, common or fairly common nesting species included veery, ovenbird, yellow-throated vireo, and red-headed woodpecker; similar species composition was found in Palmer Park (Detroit) and Highland Park. When he moved to Grosse Ile in 1907, he found that cerulean and golden-winged warblers were common nesters, as were American redstarts and orchard orioles. He frequently saw both bitterns, king and Virginia rails, and soras. The red-shouldered hawk was the most abundant nesting hawk in southeast Michigan, and vesper sparrows commonly nested in fields adjacent to city streets in Detroit. All of these species have virtually disappeared as nesting species in Wayne County (Craves, forthcoming).

Bald eagles and ospreys feeding all summer at Humbug Marsh may be nesting on both the Canadian and U.S. sides of the river. Peregrine falcons, introduced to downtown Detroit in 1988 and doing well, although native to Michigan may not be native to the area. Since 1988, other peregrines have voluntarily colonized Detroit's downtown office towers, where rock doves (domestic pigeon) abound as prey. Peregrine falcons also nest on the Canadian side of the Detroit River at the top of the Hiram Walker plant.

Impact of Exotic Species

While biodiversity is generally accepted to mean the number of organisms in a community (as well as their interconnectedness and genetic variability), the advisability of restricting that measure to native species, as suggested earlier, is readily apparent with birds. More is not always better. Several non-native bird species have been introduced to North America that have outcompeted native species and changed the composition of bird life forever. Those with the most negative impact are the European starling, mute swan, and house sparrow. All are abundant in the Detroit area.

Less well-known is that introduced non-bird species can also have an impact on bird diversity and populations. For example, the zebra mussel was introduced to Lake St. Clair in the mid-1980s. Subsequently, numbers of both lesser and greater scaup staging on the lower Great Lakes, including the Detroit River, increased as they switched to a diet dominated by the mussels (Craves 1991; Long Point Bird Observatory 2000). Zebra mussels are filter feeders and accumulate contaminants that are then passed higher up the food web.

Scaup populations have shown a significant decline since 1988; in 1998, they reached their lowest levels since surveys began in 1955 (Allen, Caithamer, and Otto 1999). Studies have indicated that when fed contaminated zebra mussels, the closely related tufted duck *(Aythya fuligula)* laid fewer, smaller eggs and productivity was reduced (de Kock and Bowmer 1993). Further studies are underway to determine if zebra mussels are contributing to the decline in scaup.

Grassy Island

Grassy Island is located off of Wyandotte. Originally it was a marshy area of emergent vegetation of six acres (2.4 ha). In one of his journals Swales describes an egg-collecting trip to Grassy Island on June 3, 1894: "This island is simply a long narrow strip of marsh entirely under water and covered with a growth of flags, marsh grass, etc. It took all the nerve I possessed to go into it. We disrobed, and the fun began. The flags cut me like a knife, and it was terrible on the feet. Besides this, the bloodsuckers, snakes, pike, etc. annoyed me dreadfully, and we never knew how far we'd sink."

He went on to describe the day's collecting bounty: black tern, 31 eggs;

common moorhen, 51 eggs; marsh wren, 50 eggs; least bittern, 4 eggs; and pied-billed grebe, 13 eggs. A return trip on June 9 found another party had preceded them and "ravaged most of the nests." Nonetheless, they made away with another 84 eggs from mostly the same species, with the addition of common yellowthroat. Similar collecting trips continued until at least 1901, with comparable results.

In 1959, the U.S. Army Corp of Engineers began diking the Grassy Island area to create a Confined Disposal Facility for dredge spoils from the Rouge River (Manny 1998). Because the areas surrounding Grassy Island contained extensive beds of wild celery, Grassy Island and environs (463 acres) were designated as the Wyandotte National Wildlife Refuge in August 1961. The U.S. Army Corps of Engineers continued to use the Confined Disposal Facility until 1983 (Sweat 1998).

The bird species found by Swales are now seldom encountered along the Detroit River. The beautiful black tern, which Swales described as nesting on the river in "immense numbers," is virtually absent as a nesting species and has so declined throughout Michigan that it is now listed as a special concern species (Michigan Natural Features Inventory 1999) as is the common moorhen. The least bittern, once common on both on Grassy Island and Grosse Ile, is now listed as threatened in Michigan.

This once gloriously productive marsh is now home to millions of cubic meters of dredged sediments sprouting weedy plants with little wildlife value (Manny 1998). Grassy Island is highly contaminated, with sediments and soils containing unacceptable concentrations of metals and organic compounds (Sweat 1998). Currently, the area is off-limits to the public (Department of Interior-Fish and Wildlife Service 1998).

The Humbug Complex

There is better news a little further south along the Detroit River's shoreline. At the southern extreme of Trenton and the northern part of Gibraltar is the Humbug Complex, the last undeveloped mile of the river's U.S. mainland shore. It includes a 30–acre island, a rich and varied marsh of about 100 acres, and an adjacent forested upland of about 320 acres. The island and mainland have varied habitat types from dense shrub borders along and grading into the coastal wetland, forests of several types, and inland vernal ponds, swamps, and marshes. Although no comprehensive survey has been done, the Humbug Marsh is known to be a migration stopover for at least 75 species of birds, including 17 raptors (Tulen et al. 1998). With a more systematic survey the numbers would surely increase. Notable among them are the threatened osprey, bald eagle, Caspian tern, common tern, and blue wing teal in numbers too large too count, and large assemblages of great blue herons and great egrets. Warblers and passerines pass through as well. Resident individuals from these same species as well as black-crowned night herons and yellow legs use the area extensively for feeding. Black-crowned night herons are of special note because the dense shrubs bordering the marsh on the island and mainland are ideal nesting and feeding habitats. There are no confirmed nesting reports, but the potential for nesting, especially since the species nests in similar habitat in Lake Erie, is great.

Conclusions

The Detroit River is an important reservoir of biodiversity, despite its industrialization and the declines from its presettlement splendor. The experiences of the wild that can still be had are astounding, and there is much at stake. What can be done to stem the tide of biodiversity loss and maybe even reverse it, right here on the Detroit River?

Although this list is by no means exhaustive, we recommend the following:

1. Pollution control: Although the return of the bald eagle and lake sturgeon among other species testifies to the progress that has been made, more needs to be done. We still have sewer overflows and the mix of chemicals, pet waste, and all manner of refuse in storm water still needs to be corralled. New U.S. Environmental Protection Agency storm-water rules should help. Mercury from power plant air emissions, which winds up in the river, is of special concern to those consuming river fish. Legislation has been proposed to severely restrict those emissions. People can do their part by adhering to environmentally friendly practices in the care of their homes, property, and boats.

2. Control of exotics: The damage done from exotics is enormous. Although it will be impossible to eradicate most species, the most important course of action now is closing the gate to new introductions. Stronger regulations and legislation on ballast water discharge would help immensely. Citizens can help with the purple loosestrife problem by volunteering to help with its eradication. Some area science classes are raising beetles, which feed only on loosestrife, for biological control.

3. Conduct species and ecosystem inventories: A comprehensive study of presettlement flora and fauna should be completed so that a more accurate assessment of the current status of biodiversity can be done. A survey and prioritization of habitats is in progress and a Detroit River biodiversity atlas is being published.

4. Education: Publishing a Detroit River biodiversity atlas similar to the one developed for Chicago will be a valuable resource for schools and citizens alike (Sullivan 1998). Instituting a K–college Detroit River Education Program similar in scope to the Rouge River Education Project would go a long way to help people in the community understand what an incredible resource the river is and how they can help protect and improve it. Work projects such as native habitat restoration on Belle Isle should continue.

5. Integrate conservation into regional planning: This has to become part of our everyday activities if it really is to make a long-term difference.

6. Sustainable development: Undertaking development such that the use of natural resources is not increased and environmental impacts are minimized is the only way to preserve biodiversity. Locally, Sisters of the Immaculate Heart of Mary are pioneers in employing this process and commitment in the development of their 280–acre property in Monroe, reusing materials, recycling others, and involving the community so that multiple needs can be met. A good example of the use of a remodeled building is the National Audubon Society's headquarters in New York City.

7. Preserving and restoring habitat: Since loss of habitat has been the biggest factor in biodiversity decline, it is critical that what we have left is protected, which includes not only existing parks and reserves but the Humbug Complex and Sibley Prairie as well. In addition to preserving relatively intact reservoirs of biodiversity (refugia), restoration is needed as well. Restoration or re-creation of the following ecosystems would help enhance the Detroit River's ability to maintain and perhaps increase the diversity of native species:

- coastal marshes on Belle Isle and on the U.S. and Canadian mainland wherever possible (their usefulness would be enhanced if they graded into shrubs and then into existing or restored native forests)
- prairies on Belle Isle and on vacant land on Detroit's east side

- common tern colonies on Belle Isle and Bob-Lo Islands
- low grassland habitats where there is sufficient vacant land (these areas may be disappearing faster than wetlands, and grassland birds and animals along with them)

There is also a need to rehabilitate forests that have been degraded and consider allowing some vacant areas to succeed into native forests. Management will be needed to control weedy species. In addition, efforts are needed to halt drainage of wetlands and restore soft shorelines and flood regimes.

8. Involve people in diverse projects to help preserve and enhance biodiversity: The projects themselves do a lot of good, but the most important result of such projects is public awareness and involvement in conservation.
9. Reintroduce species where appropriate: For example, if shallow coastal marshes were re-created along the river's shoreline, with at least a partial sand substrate, the experience at Metzger Marsh on Lake Erie suggests that such areas can be refugia for native mussels (U.S. Geological Survey 1996).

As we reflect on the Detroit River's glorious past and the mixed blessings of the present we might do well to keep in mind these words of a well-known conservation biologist: "Some people feel discouraged by the avalanche of species extinctions occurring in the world today, but it is also possible to feel challenged by the need to do something to stop the destruction" (Primack 1995, 2). It is up to us to recognize that challenge and meet it.

LITERATURE CITED

Allen, G. T., D. F. Caithamer, and M. Otto. 1999. "A Review of the Status of Greater and Lesser Scaup in North America." U.S. Fish and Wildlife Service, Office of Migratory Bird Management.

Bird Studies Canada. 1999. *Canadian Important Bird Area, Lower Detroit River. http://www.bsc-eoc.org.*

Blokpoel, H., and G. D. Tessier. 1986. "The Ring-billed Gull in Ontario: A Review of a New Problem Species." Occasional Paper No. 57. Canadian Wildlife Service, Burlington, Ont.

"Birds to Keep Their Belle Isle Roost." 1964. *Detroit Free Press,* June 19, p. 3A.

Brewer, R., G. A. McPeek, and R. J. Adams, Jr., eds. 1991. *The Atlas of the Breeding Birds of Michigan.* East Lansing: Michigan State University Press.

Cook, A. J. 1893. *Birds of Michigan.* Michigan Agricultural Experiment Station Bulletin 94.

Cooper, G. P. 1952. *The Fish Fauna and the Fishing of the Detroit River in the Vicinity of Sugar and Stony Islands.* Michigan Department of Conservation Report #1350. Ann Arbor, MI.

Craves, J. A. 1991. "Zebra Mussels and Bird Life: Is There a Connection?" *Nuthatch, Newsletter of the Oakland County, Michigan Audubon Society* (summer 1991): 6.

———. 1996. *The Birds of Southeast Michigan: Dearborn, Wayne County.* Bloomfield Hills, MI: Cranbrook Institute of Science.

———. Forthcoming. "Historical Changes in Breeding Bird Populations of Wayne County, Michigan."

Craves, J. A., and O. G. Gelderloos. 1996. "Birds of the Rouge River Floodplain, Dearborn, Michigan: Importance of an Urban Natural Area to Birds." *Michigan Birds & Natural History* 3:3–12.

de Kock, W. C., and C. T. Bowmer. 1993. "Bioaccumulation, Biological Effects, and Food Chain Transfer of Contaminants in the Zebra Mussel *(Dreissena polymorpha)."* In *Zebra Mussels: Biology, Impacts, and Control,* ed. T. F. Nalepa and D. W. Schloesser, 503–36. Boca Raton: Lewis Publishers.

Department of Interior-Fish and Wildlife Service. 1998. *Notice of Intent to Prepare Comprehensive Conservation Plans and Associated Environmental Documents; Michigan and Minnesota.* Federal Register 63.247: December 24, 1998. *http://www .eswr.com.*

Dowdy, E. B. 1964. "Waterfowl Friends, Foes Joust on Council Floor: Try to Stop Belle Isle Nesting." Detroit News, June 19, p. 2A.

Farmer, S. 1890. *History of Detroit and Wayne County and Early Michigan: A Chronological Cyclopedia of the Past and Present.* Detroit: Silas Farmer and Company.

Galli, A. E., C. F. Leck, and R.T.T. Forman. 1976. "Avian Distribution Patterns in Forest Islands of Different Sizes in Central New Jersey." *Auk* 93:356–64.

Gleason, H. A., and A. C. Cronquist. 1963. *Manual of Vascular Plants of Northeastern United States and Adjacent Canada.* New York: Van Nostrand Reinhold Company.

Haas, R. C., W. C. Bryant, K. D. Smith, and A. J. Nuhfer. 1985. *Movement and Harvest of Fish in Lake St. Clair, and Detroit River: Final Report of Winter Navigation Study.* Detroit: U.S. Army Corps of Engineers.

Hartig, J. H. 1993. "A Survey of Fish-Community and Habitat Goals/Objectives/Targets and Status in Great Lakes Areas of Concern." Ann Arbor, MI: Great Lakes Fisheries Commission.

Hobbs, C. L., and K. F. Lagler. 1958. *Fishes of the Great Lakes Region.* Ann Arbor: University of Michigan Press.

Jeremy, Long, Sands, Stork, and Winser, eds. 1995. *Biodiversity Assessment: A Guide to Good Practice.* London: Department of the Environment (HSMO).

Lanigan, J. 1998. "BASF Corporation's rehabilitation of Fighting Island." In *Rehabilitating and Conserving Detroit River Habitats,* ed. L. A. Tulen, J. H. Hartig, D. M. Dolan, and J. Ciborowski, 36–38. Winsdor, Ont.: Great Lakes Institute for Environmental Research.

Lapin, B., and J. Randall. 1993. *Element Stewardship Abstract:* Phragmites australis *(Common Reed). http://theweeds.ucdavis.edu/esadocs/documents/phraaus.html*

Long Point Bird Observatory. 2000. "LPWWRF Starts Two New Research Projects." *Long Point Bird Observatory and Ontario Programs Newsletter* 32.2:9–11.

Ludwig, J. P. 1962. "A Survey of Gull and Tern Populations of Lakes Huron, Michigan, and Superior." *Jack-Pine Warbler* 40:104–19.

———. 1991. "Ring-billed Gull." In *The Atlas of the Breeding Birds of Michigan,* ed. R. Brewer, G. A. McPeek, and R. J. Adams, Jr., 216–17. East Lansing: Michigan State University Press.

Manny B. A. 1998. "Ecological Restoration of Grassy Island and the Wyandotte National Wildlife Refuge in the Detroit River." In *Rehabilitating and Conserving Detroit River Habitats,* ed. L. A. Tulen, J. H. Hartig, D. M. Dolan, and J. Ciborowski, 18–21. Windsor, Ont.: Great Lakes Institute for Environmental Research.

Manny, B. A., T. A. Edsall, and E. Jaworski. 1988. *The Detroit River: An Ecological Profile.* U.S. Fish and Wildlife Service Biological Report 85 (7.17). Ann Arbor, MI.

Michigan Department of Natural Resources. 1998. *Endangered, Threatened and Special Concern Species. http://www.dnr.state.mi.us.*

Michigan Natural Features Inventory. 1999. *Michigan's Special Animals. http://www .michigan.gov/dnr*

Miller, H. J. 1943. *Waterfowl Survey, Saginaw Bay-Lake St. Clair- Lake Erie.* Project N, 13-R. Lansing: Michigan Department of Natural Resources, Wildlife Division.

Moore, F. R., and T. R. Simons. 1989. "Habitat Suitability and Stopover Ecology of Neotropical Landbird Migrants." In *Ecology and Conservation of Neotropical Migrant Landbirds,* ed. J. M. Hagan, III and D. W. Johnston, 345–55. Washington, DC: Smithsonian Institute Press.

Moore, F. R., S. A. Gauthreaux, Jr., P. Kerlinger, and T. R. Simons. 1995. "Habitat Requirements during Migration: Important Link in Conservation." In *Ecology and Management of Neotropical Migrant Birds: A Synthesis and Review of Critical Issues,* ed. T. E. Martin and D. M. Finch, 121–44. New York: Oxford University Press.

Muller, W. 1964. "Tern Justice: Water Intake Issue." Detroit News, June 16, p. 8B.

Muth, K. M., D. R. Wolfert, and M. T. Bur. 1986. *Environmental Study of Fish Spawning and Nursery Areas in the St.Clair-Detroit River System.* Sandusky, OH: U.S. Fish and Wildlife Service, Sandusky Biological Station.

Primack, R. B. 1995. *A Primer of Conservation Biology.* Sunderland, MA: Sinauer Associates Inc.

Reid, R., K. Rodriguez, and A. Mysz. 1999. *State of the Lakes Ecosystem Conference: Biodiversity Investment Areas—Nearshore Terrestrial Ecosystems.* Chicago and Burlington, Ont.: U.S. Environmental Protection Agency and Environment Canada.

Rouge River Bird Observatory. 2000. *Dearborn Bird Checklist. http://www.umd .umich.edu/rouge-river/rr-birds.html*

Scharf, W. C. 1991. "Species Account, Common Tern *(Sterna hirundo)*." In *The Atlas of Breeding Birds of Michigan,* ed. R. Brewer, G. A. McPeek, and R. J. Adams, Jr., 222–23. East Lansing: Michigan State University Press.

Schloesser, D. W., and B. A. Manny. 1990. "Decline of Wild Celery Beds in the Lower Detroit River, 1950–85." *Journal of Wildlife Management* 54:72–76.

Schloesser, D. W., W. P. Kovalak, G. D. Longton, K. L. Ohnesorg, and R. D. Smithee.

1998. "Impact of Zebra and Quagga Mussels *(Dreissena spp.)* on Freshwater Unionids (Bivalva: Unionidae) in the Detroit River of the Great Lakes." *American Midland Naturalist* 140:299–313.

Schwartz, M. W. 1997. *Conservation in Highly Fragmented Landscapes.* New York: Chapman and Hall, International Thomson Publishing.

Southeast Michigan Council of Governments (SEMCOG). 1999. *A Profile of Southeast Michigan's Environment.* Detroit: SEMCOG

Smith, H. M. 1917. *Report of the Bureau of Fisheries. Report on the U.S. Commission of Fisheries for 1916.* Ann Arbor, MI: U.S. Geological Survey-Great Lakes Science Center.

Smith, R. P. 1997. *Our "Downriver" River: Nautical History and Tales of the Lower Detroit River.* Detroit: Rockne P. Smith.

Sullivan, J. 1998. *Chicago Wilderness: An Atlas of Biodiversity.* Chicago: Chicago Region Biodiversity Council.

Swales, B. H. Journals, 1893–1915. University of Michigan Museum of Zoology, Bird Division, Ann Arbor.

Sweat, M. J. 1998. *Investigation of Sediment and Water Chemistry at Grassy Island, Shiawassee National Wildlife Refuge, Wyandotte Unit, Wyandotte Michigan.* http://mi.water.usgs.gov.

Tessen, D. D. 2000. "Western Great Lakes Region: The Changing Seasons, Fall Migration, August–November 1999." *North American Birds* 54:52–57.

Tulen, L. A., J. H. Hartig, D. M. Dolan, J. J. H. Ciborowski, eds. 1998. *Rehabilitating and Conserving Detroit River Habitats.* Windsor, Ont.: Great Lakes Institute for Environmental Research.

University of Minnesota, Department of Fisheries and Wildlife. 2000. *Biological Control of Eurasian Watermilfoil.* http://www.fw.umn.edu.

University of Minnesota Sea Grant. 2000. *Field Guide to Aquatic Exotic Plants and Animals.* http://www.d.umn.edu.

U.S. Fish and Wildlife Service. 2000. *Aquatic Nuisance Species.* http://www.midwest.fws.gov/Alpena/index.html

———.1998. September 30th Letter to U.S. Army Corps of Engineers regarding development of Humbug Marsh (Joint Public Notice 88-007-79-4D/98-10-328)

U.S. Geological Survey. 1996. *Coexistence of Zebra Mussels and Clams in a Lake Erie Watershed.* Great Lakes Science Center Fact Sheet 96-9. Ann Arbor, MI.

Wood, J. C. 1909. "Warbler Notes from Wayne County, Michigan." *Wilson Bull* 21:45–46.

Yong, W., and F. R. Moore. 1997. "Spring Stopover of Intercontinental Migratory Thrushes along the Northern Coast of the Gulf of Mexico." *Auk* 114:263–78.

Yong, W., D. Finch, F. R. Moore, and J. F. Kelly. 1998. "Stopover Ecology and Habitat Use of Migratory Wilson's Warblers." *Auk* 115:829–42.

PREVENTING TOXIC SUBSTANCE PROBLEMS THROUGH DESIGN AND INDUSTRIAL CONTROL OF CONTAMINANTS AT THEIR SOURCE

D. C. Steinmetz and D. P. Thiel

The Detroit River has for thousands of years been a thirty-two-mile connecting channel between Lake St. Clair and Lake Erie. Prior to the 1800s, a flat plain that stretched out for over fifteen miles surrounded the river. Several small rivers and many creeks, which fed into the Detroit River, traversed this flat plain. In addition, there were many shallow bays that were also adjacent to the river (Woodford and Woodford 1969). This ecosystem was very attractive to fish and wildlife, which inhabited this area in large numbers. Over the centuries, the uniqueness of this river ecosystem, in turn, attracted various human populations that lived off of the bounty in and along the river. These human populations had a minor impact on the water quality of the Detroit River ecosystem as their activities involved largely hunting, fishing, and agriculture.

It was not until the mid-1800s that a significant human population developed along the Detroit River with the infusion of people from Europe. It is estimated that in 1850 there were approximately 20,000 people living in the city of Detroit. At that time the city's population growth started to accelerate. To accommodate the growing number of people, sailing ships brought lumber and other goods to the city from up north. This led to a construction boom and by the late 1800s Detroit became the busiest port in the world. Farming, lumber, and small industry dominated the local economy at that time.

Early in the 1900s, heavy industry started to crop up along the river. Among the early industries were automobile, barrel, chemical, and steel manufacturing. Other operations included forging, shipbuilding, salt mining, silver smelting, and tanning (DeWindt and DeWindt 1955). This major industrial growth led to many uncontrolled discharges into the Detroit River. During a time when belching smokestacks were seen as a sign of industrial progress and might, industry and the general public either chose to ignore or did not realize the overall adverse impact this would have on the Detroit River ecosystem. The collapse of the commercial fishing industry in the Detroit River at the close of the nineteenth century (DeWindt and DeWindt 1955) was a strong indication that something was amiss; however, this sign was largely ignored. In 1913, a typhoid epidemic occurred in Detroit, which resulted from a contaminated drinking water supply. It caused 153 deaths and the response was to relocate the intake for Detroit's drinking water supply to avoid the source of the problem (chapter 5). Industrial and municipal discharges continued unabated for many decades until a number of major water pollution incidents occurred that highlighted the escalating adverse ecological impacts. These pollution incidents included:

- the 1948 winter duck kill on the Detroit River, where 11,000 ducks died due to oil pollution (chapter 6)
- eutrophication of Lake Erie during the 1960s due to phosphorus pollution from detergents and municipal wastewater (chapter 7)
- the 1970 mercury crisis in Lake St. Clair (chapter 8)

Pollution of the Detroit River could not be ignored any longer. The general attitude toward pollution of the river has evolved from denial, to control, through prevention, and now toward design for sustainability (table 12.1).

Historical Water Pollution Abatement Approaches

The major water pollution incidents led to a focus on "end-of-pipe" or "command-and-control" regulations aimed at controlling pollution at the end of a pipe or stack. The first significant piece of legislation, which focused on the command-and-control regulation of pollutants, was the Clean Water Act enacted in 1972. The act provided the basic national framework for water pollution control and water quality management in the United States. The U.S. Environmental Protection Agency (EPA) has nationwide authority to implement this act. Amendments to the act in 1977 and 1987 strengthened this framework to protect the nation's waters.

TABLE 12.1

Summary of Changes in the General Attitude Toward Pollution and the Focus of Environmental Management Efforts

Time Period	Attitude and Focus
1900–50	Denying, ignoring, or avoiding pollution
1950–mid-1990s	Reducing pollution through end-of-pipe regulations and controls
Mid-1990s–2000	Preventing pollution and ensuring product stewardship (i.e., minimizing the pollution associated with the entire life cycle of a product from extraction of raw materials to manufacturing, use, recycling, and disposal)
2000–	Designing products/processes and developing clean technologies for environmental, economic, and societal sustainability

To provide additional protection to the waters within their boundaries, a number of states enforce state water laws providing even greater environmental protection than the federal Clean Water Act. In adopting the Clean Water Act, Congress recognized that the states could manage water quality control more effectively than the EPA could. As a result, the Clean Water Act specifically provides that states with a water quality program as environmentally protective as the Clean Water Act may apply to the EPA for authorization to administer various aspects of the Clean Water Act's program for permitting discharges into waterways. This is the case in Michigan, where compliance with the state's Natural Resources and Environmental Protection Act (the state act) satisfies the regulatory requirements of the Clean Water Act and allows the state to impose water quality controls over the surface waters in the state, including the Detroit River. The Michigan Department of Environmental Quality, Surface Water Quality Division, administers the state act.

The basis of water quality control under the Clean Water Act and the state act is two-pronged. One focus is the control of pollutant discharges into the nation's waters from primarily industrial "point sources." In regulating these discharges, the Clean Water Act controls the types and amounts of pollutants in specific industries' waste streams. These controls are known as "technology-based effluent limitations." The other major thrust of the regulations is the requirement for the states to categorize navigable waters within their boundaries according to use (e.g., recreational, agricultural, industrial) and establish water quality standards appropriate to the ascribed uses. The purpose is to ensure the quality of the water for these designated uses and to restrict discharges that will degrade the water source below water quality standards. This approach is known as water quality-based control.

The Clean Water Act ensures that water quality standards for waters receiving discharges are met and that each discharger applies the required technology-based standards to the facility through the application of a comprehensive permitting program known as the National Pollutant Discharge Elimination System (NPDES). The NPDES program requires the owner/operator of a facility to obtain a permit prior to discharging any pollutant into the navigable waters of the United States from any point source. Examples of point sources would be a pipe, ditch, channel, well, or vessel. Therefore, the focus of water pollution control over the past fifty years or so has been on end-of-pipe controls and application of increasingly restrictive limits on polluting materials.

This strategy made both environmental and economic sense during this era because of the higher pollution levels. However, once basic technological requirements were established and implemented for specific industrial and municipal discharges, this strategy produced an inverse relationship between the resources allocated and the environmental benefit as more complex and expensive controls were required to capture or control lower concentrations of polluting substances. The transactional costs and adversarial relationships associated with administering, permitting, and enforcing such a system rose substantially over time.

Recent Approaches

Through the first half of the twentieth century, industry along the Detroit River, like industries around the globe, looked at surface waters as a source of process and cooling water and as a mechanism for diluting and carrying away contaminants. With the advent of federal, state, and local legislation, and a higher level of concern for the environment brought about by publications such as Rachel Carson's *Silent Spring* in 1962, industries began to develop programs and processes to protect the quality and safety of waterways. The activities were generally aimed at minimizing waste and preventing pollution. In general, pollution prevention means preventing or reducing waste where it originates, at the source, including practices that conserve natural resources by reducing or eliminating pollutants through increased efficiency in the use of raw materials, energy, water, and land.

At the 1992 Earth Summit in Rio de Janeiro, the concerns associated with rapid worldwide economic development and explosive population growth were galvanized into the concept of sustainable development. The World Commission on Environment and Development defined sustainable development as "development that meets the needs of the present without compromising the ability of future generations to meet their needs." Sustainability became a major catalyst for change.

Over the last decade, the focus of many industries along the Detroit River and around the world has shifted from pollution prevention to the prevention of waste through the development of sustainable products and processes. End-of-pipe controls will become less pervasive as companies begin to respond to new market drivers for these new eco-efficient products.

While many companies are striving to use fewer resources to accomplish the same objectives, others are beginning to recognize that product and process design changes early in the life cycle can be more effective in avoiding the depletion or use of critical raw materials. Life-cycle assessment and management tools are being developed to determine the true environmental costs of producing materials and products across the value chain. The real costs and impacts on the environment are not always evident when viewed from the perspective of one person using it for part of its life. The cumulative impacts of disposal, manufacturing, reuse, and recycling are critical in determining a product's true "cost" to the environment so that industries and consumers can make intelligent choices related to their activities.

These life-cycle assessment and management tools have been integrated into the concept of "Design for Environment," which is the systematic consideration, during product design, of issues associated with environmental safety and health over its entire life cycle. "Design for Environment" can be thought of as the migration of traditional pollution prevention concepts upstream into the development phase of products before production and use. "Design for Environment" is being applied to the design of new and modification of existing products, processes, and facilities. The objective is to minimize or eliminate, during design, the anticipated waste generation, resource consumption, and environmental burden in all subsequent life-cycle phases (e.g., resource extraction, production, use, and disposal).

The concept of eco-efficiency is essentially a strategy for developing industrial systems and processes that deliver competitively priced goods and services to satisfy human needs and achieve quality of life, while progressively reducing ecological impacts and resource intensity to a level in line with earth's estimated carrying capacity (DeSimone and Popoff 1997; Fussler and

James 1996; Lerner 1997). This philosophy of "doing more with less" has roots in early industrialization. Henry Ford, for example, was a strong proponent of saving his company money through the recycling and reuse of waste.

Chemical, steel, and other manufacturing companies along the Detroit River have had success in reducing waste and emissions to the air and water and have learned that end-of-pipe control methods are often costly and inefficient. Many chemical companies doing business along the river support and practice the principles of Responsible Care®. These principles include:

- to seek and incorporate public input regarding our products and operations
- to provide chemicals that can be manufactured, transported, used, and disposed of safely
- to make health, safety, the environment, and resource conservation critical considerations for all new and existing products and processes
- to provide information on health or environmental risks and pursue protective measures for employees, the public, and other key stakeholders
- to work with customers, carriers, suppliers, distributors, and contractors to foster the safe use, transport, and disposal of chemicals
- to operate facilities in a manner that protects the environment and the health and safety of our employees and the public
- to support education and research on the health, safety, and environmental effects of our products and processes
- to work with others to resolve problems associated with past handling and disposal practices
- to lead in the development of responsible laws, regulations, and standards that safeguard the community, workplace, and environment
- to practice Responsible Care® by encouraging and assisting others to adhere to these principles and practices

Responsible Care® is a voluntary initiative within the global chemical industry to safely handle products from inception in the research laboratory, through manufacture and distribution, to ultimate disposal, and to involve the public in decision-making processes. It can best be described as a series of management practices designed to promote responsible production and handling of chemicals. Begun in Canada in 1987, Responsible Care® has quickly spread to forty-five countries. This international effort has contributed to significant reductions of emissions to air, water, and land. Many companies have altered their processes to eliminate the use of potentially toxic materials. Reductions in "toxic releases" have been noted by many of the industries lining the river as well as in Michigan overall over the past twelve years (table 12.2). However, further reductions will be needed to ensure environmental protection.

The Michigan Department of Environmental Quality (MDEQ) is also developing programs and providing incentives to industry to help reduce the environmental impacts of their operations. For example, the Michigan Business Pollution Prevention Partnership (MBP3) is a voluntary program, open to the entire business community. It was established to encourage businesses to initiate and/or expand pollution prevention (P2) practices and provides recognition for these efforts. The MBP3 encourages businesses to apply innovative and cost-effective techniques to prevent releases of hazardous substances and to reduce waste. This program was developed through a joint effort by business, industry, professional associations, and the MDEQ.

Pollution prevention has as a first preference the use of source reduction. Pollution prevention also includes as a second preference reuse and environmentally sound recycling techniques. Pollution prevention avoids cross-media transfer of waste and/or pollutants. It addresses all types of waste and environmental releases to air, water, and land.

Pollution prevention offers important economic benefits and at the same time allows continued protection of the environment. While most pollution control strategies cost money, pollution prevention has saved many firms thousands of dollars in raw materials, waste treatment, and disposal costs. For example, a Saline, Michigan, book printer instituted a pollution prevention program and reduced the company's solid wastes by 50 percent. In 1995, the program recovered $280,000 by recycling film plates and paper. Another Michigan firm increased its adhesive usage efficiency, thus reducing the gallons of adhesive purchased, saving approximately $500,000 annually in raw material costs.

To participate in this program, businesses must submit a statement of commitment signed by a responsible company official. Participating businesses must adopt a pollution prevention policy, identify and commit to prevention and reduction goals, report on goal achievement and other pollution prevention accomplishments, and share technological success stories that can be transferred to other business operations.

Low-interest loans have recently become available through the MDEQ to small businesses to improve their efficiency and reduce waste. The loans may be up to $100,000 at a 5 percent or less interest rate. They must be used

TABLE 12.2

Summary of Changes in Reported Toxic Release Inventory Data from the Environmental Defense Scorecard for the State of Michigan, 1988–2001 (Environmental Defense 2002; Toxic Release Inventory [TRI] Data Summary, Environmental Releases, Transfers, and Production-related Waste for Michigan TRI Reporting Facilities)

Year	Air Releases	Water Releases	Land Releases	Underground Injection	Total Environmental Releases	Total Offsite Transfers	Total Production-Related Waste
1988	101,148,925	1,222,723	6,652,308	6,489,618	115,513,574	155,597,955	N/A
1989	105,071,755	738,712	9,192,744	8,278,831	123,282,042	95,416,025	N/A
1990	84,349,537	818,195	9,256,179	8,323,247	102,747,158	72,654,624	N/A
1991	72,509,900	957,641	13,949,308	6,699,997	94,116,846	233,775,200	863,836,187
1992	68,153,224	762,404	10,328,826	6,083,782	85,328,236	256,369,262	976,455,309
1993	69,247,799	584,065	9,137,340	5,619,980	84,589,184	314,444,423	1,014,330,059
1994	64,937,744	1,000,017	9,719,406	6,962,868	82,620,035	286,561,233	1,025,496,410
1995	62,996,379	653,999	4,046,748	7,566,827	75,263,953	239,922,589	739,423,353
1996	46,131,783	1,989,436	3,654,951	6,632,503	58,408,673	239,548,311	799,753,252
1997	44,323,923	529,145	2,595,399	5,596,855	53,045,322	493,214,203	866,066,923
1998	75,069,608	685,649	19,957,839	3,431,059	99,144,155	442,020,350	1,200,158,636
1999	76,110,644	1,265,578	17,187,904	2,571,942	97,136,068	362,592,667	1,075,566,377
2000	71,193,709	1,134,859	33,039,616	2,057,816	107,426,001	304,379,560	662,004,359

Note: "N/A" means that no data are available because "Total Production-Related Waste" was not reported until 1991.

to implement a project that eliminates or reduces waste through pollution prevention. The MDEQ will provide a certificate to all business program participants. An annual Michigan Pollution Prevention Report is issued by the MDEQ summarizing the previous year's accomplishments.

The Clean Corporate Citizen (C3) Program, administered through the MDEQ, allows regulated establishments that have demonstrated strong environmental performance and a commitment to continual environmental improvement to be recognized as Clean Corporate Citizens. Qualifying businesses are eligible for benefits including expedited permit reviews, fewer monitoring and reporting requirements, and a more positive public perception, which can add value to a company's competitive position in today's global marketplace. This is a voluntary program that has attracted a wide range of establishments including auto assembly, power generation, furniture making, natural gas transmission, paper production, office management, research and development, and barrel reconditioning.

In order to participate, a company must have in place a strong and effective environmental management system, pollution prevention program, and a stellar environmental compliance record with no outstanding unresolved violations. The actual application process begins when the applicant provides public notice and a forum for public review of a completed application and supporting materials. After a thirty-day public review, the candidate firm submits the application and supporting materials to the MDEQ for review and subsequent approval or denial of the C3 designation. The designation can be renewed annually if the establishment continues to meet the C3 criteria. As of 2002, there have been forty-eight facilities that have earned this designation in Michigan.

Leadership for Sustainability

General Motors Corporation, whose global world headquarters is located on the Detroit River and who has a number of plants in the watershed, is a world leader in transportation products and related services and in practical application of sustainable development. General Motors' view of sustainability is captured in the following quote from Vice Chairman Harry Pearce: "We are all responsible for this planet, but business must take the lead, because only business has the global reach, the innovative capability, the capital, and most importantly, the market motivation to develop the technologies that will allow the world to truly achieve sustainable development." Sustainable development is a management framework that drives continuous improvement in General Motors' daily business and allows it to view challenging issues as new business opportunities. To be able to help measure progress, General Motors set a goal of reducing energy use in North America by 20 percent by 2002 from 1995 levels and to reduce emissions to air, land, and water by 30 percent from 1997 levels.

General Motors has developed a set of sustainable development indicators, which provides estimates of energy used, waste emitted, and water used to produce an automobile. These indicators follow the highly respected Global Reporting Initiative's Sustainability Reporting Guidelines. In a comprehensive report titled *Steps toward Sustainability,* General Motors (2000) reported a 24 percent reduction in emissions from North American plants in 1998 and 1999 (this was due to a dramatic reduction of zinc emissions from foundries; zinc emissions declined from a high of 6,350 metric tons per year in 1995 to 284 metric tons in 1999). It is metrics like these that will ultimately provide a way for the public to measure the environmental impact of producing the products and services that we consume.

"Design for Environment" is being used by each of General Motors' new product teams. For example, the General Motors Truck Group successfully factored environmental considerations into the 1999 Chevrolet Silverado and GMC Sierra. There are several benefits and achievements of this approach:

- Elimination of mercury from underhood lamp switches avoids the potential release of more than one ton of mercury per model year.
- Increased fuel economy translates to a savings of 3.7 million barrels of oil or over 157 million gallons of gas over the life of one model year's production.
- Replacement of lead solder with a new spliceless design in the power and signal distribution areas eliminates 18 tons of lead per model year.
- Increased brake life expectancy of more than twofold reduces scrapped brake parts by more than 6,000 tons per model year.
- Production of wheel cap assemblies from recycled regrind scrap from Saturn fenders makes use of scrap and saves money.
- Use of new hydroforming frame technologies reduces steel scrap by 19,800 tons per model year.
- Recycling of old tires to produce radiator side air baffles uses 140,000 tires per model year.
- Painting of grills, mirror housings, and bumper caps has been eliminated.
- The overall vehicles are 89 percent recyclable

Figure 12.1. The new Ford Rouge Complex that will pioneer and model sustainable manufacturing and productive landscape (Richard Rochon; William McDonough and Partners). This reflects William Clay Ford, Jr.'s vision for sustainable redevelopment of the Rouge industrial complex (note grass on buildings and vines on sides). (Wayne County Rouge River National Wet Weather Demonstration Project)

General Motors also believes that advanced technologies are the key to continued improvements in fuel economy and reductions in carbon dioxide from cars and light trucks. Since 1996, General Motors has had the EV1, the most technologically advanced electric vehicle in the market. The Chevrolet S10 electric pickup truck has been available for commercial and government fleets since 1997. The advanced nickel-metal hydride batteries and a Generation II electric drive were introduced in 1998 for the EV1 and electric S10 pickup that significantly increased the driving range for these vehicles. General Motors believes such advanced technologies hold great promise and will lead to competitive advantage.

Decades after Henry Ford promoted the practice of recycling, William Clay Ford, Jr., after being named chairman of Ford Motor Company, announced a project to remake the enormous Rouge industrial complex into a model of sustainable production (figure 12.1). Originally spanning more than 1,100 acres, the Ford Rouge industrial complex was the largest industrial complex in the world when it began producing the Model A in 1927. But the once innovative method of converting raw materials into automobiles at a single complex caused considerable environmental damage over the decades. Ford Motor Company will be investing $2 billion in renovating the Rouge Complex, updating the factories, and restoring the environment. It will literally be a model of sustainable redevelopment.

Buildings that historically faced inland will in the future be facing the Rouge River. A system of swales or landscaped ditches, along with retention ponds, will be used to treat storm water. Native plants will be used to clean contaminated soils, while renewable energy sources such as fuel cells and solar cells will be incorporated into the complex. Grass on some building roofs will help control rainwater runoff, while vines will act as insulation.

Through the use of "eco-effective" products and processes, Ford will strive to make the site a model for environmentally friendly industrial production. The company will work with suppliers to "design out" emissions and wastes and revitalize the area as a "green" place for the community (table 12.1). Biodiversity in the Rouge River and other water quality measures will be employed to chart the success of the project. Plans also include trails linking to Greenfield Village/Henry Ford Museum and the Detroit River. Boat docks, observation towers, and soft engineering of shorelines (to enhance habitat) will be added to enhance tourism, the second largest industry in Michigan. Ford also plans to reopen the Rouge Complex to public tours that attracted two million annual visitors during the 1960s.

Other local companies are also looking forward. DTE Energy, which operates several large energy generation facilities across southeastern Michigan, is committed to many voluntary environmental programs. DTE, in recognizing that it "borrows" water from local lakes and rivers, including the Detroit River, manages programs to use and discharge the water in an environmentally responsible manner. This commitment extends to air emissions as well; the company reports that for the period 1974–98 it reduced air emissions of particulates by 90 percent while increasing fossil fuel power generation by 22 percent. Although DTE has made strides in reducing emissions, it recognizes that simply reducing waste will not be sufficient in providing a sustainable future. Like other companies, DTE is beginning to explore new ideas and technologies for creating sustainable processes and products. It is committing resources to developing alternative energy sources that will power future generations' needs while further reducing or even eliminating their impact on precious natural resources like the Detroit River. It is through the development of these new technologies that we may begin to realize truly sustainable energy production (figure 12.2).

The chemical industry, a significant presence along the Detroit River, is also committing substantial resources to reducing waste and emissions. Over the last decade, air and water emissions into the river watershed have decreased dramatically from many companies. More important, the industry

Figure 12.2. Wind turbines from Altamont Pass in the San Francisco Bay Area that can provide sustainable energy production (Bay Area Photo Gallery).

is beginning to focus on the development of products using "green chemistry," which seeks to decrease the environmental footprint of chemicals at the design phase. Over time, products made from these new processes will significantly reduce the environmental impact during the production phase, as well as the impact consumers cause during use. Already, companies are applying the results of green chemistry into manufacturing products such as ibuprofen, recycling and regenerating waste plastics into new materials for reuse, and developing more environmentally responsible ways of coating automobiles.

The Future

During the latter part of the twentieth century, industry and society began to understand the true monetary, environmental, and social costs of our failure to maintain the quality of the lakes, streams, and rivers that provide water for drinking, agriculture, recreation, and ecosystems. Looking forward, society and governments will increasingly demand that products and services reflect true environmental costs. The need to find competitive advantage will force industries to design new products and clean technologies to fulfill society's demands while allowing for sustainable production (Hart 1997). New energy sources will allow us to fulfill our power requirements without releasing pollutants to the air and water, better designed products will be created from raw materials that are renewable or recyclable, and industrial processes will be redesigned to reduce or eliminate pollutants. As societal demands increase, only those companies that can demonstrate sustainable production will survive.

William McDonough is an internationally renowned architect and industrial designer. He is a revolutionary thinker who offers ways in which industry and society can be redesigned to provide a positive future for the next generation. According to this 1996 recipient of the Presidential Award for Sustainable Development:

> There are millions of difficult challenges and delightful opportunities ahead. I think the only constraint is the willingness to dream, to create and to hope and feel undefended enough to face the tough questions and the clash of ideas that must be fiercely engaged at this moment in human history. If design is the signal of human intention then we must continually ask ourselves—What are our intentions for our children, for the children of all species, for all time? How do we profitably and boldly manifest the best of those intentions?

LITERATURE CITED

DeSimone, L., and F. Popoff. 1997. *Eco-Efficiency: The Business Link to Sustainable Development.* Cambridge, MA: MIT Press.

DeWindt, E., and J. DeWindt. 1955. *Our Fame and Fortune in Wyandotte.* Wyandotte, MI: Wyandotte Rotary Club.

Environmental Defense. 2002 Environmental Defense Scorecard. New York, NY. *http://www.scorecard.org.*

Fussler, C., and P. James. 1996. *Driving Eco Innovation.* London: Pitman.

General Motors. 2000. *Steps toward Sustainability: General Motors 1999–2000 Report on Economic, Environmental, and Social Performance. http://www.gm.com.*

Hart, S. L. 1997. "Beyond Greening: Strategies for a Sustainable World." *Harvard Business Review* 75.1:66–76.

Lerner, S. 1997. *Eco-Pioneers.* Cambridge, MA: MIT Press.

Woodford, F., and A. Woodford. 1969. *All Our Yesterdays.* Detroit: Wayne State University Press.

Watershed Planning and Management: The Rouge River Experience

Catherine J. Bean, Noel Mullett, Jr., and John H. Hartig

There were many warm summer nights in southeast Michigan during the summer of 1983. What made this summer so different from other ones was that residents living near the Rouge River in the cities of Melvindale and Dearborn could not keep their windows open at night. If they did, the smell of rotten eggs would permeate their living rooms and bedrooms. The smell was atrocious. What was causing this? What could be done to get rid of this unacceptable smell?

People living in Melvindale and Dearborn quickly figured out that the smell was coming from the Rouge River. Most people had heard that the Rouge River was polluted, but this smell was too much to put up with. The problem with the Rouge River was that it was not polluted with rotten eggs but with so much untreated sewage and organic waste from sanitary sewer overflows, combined sewer overflows (where storm and sanitary sewers are combined and result in the discharge of raw sewage during rainfall events), and illegal discharges that no oxygen was left in the river. Natural decomposition of raw sewage and organic waste was using up all the oxygen in the river. The smell was actually the out-gassing of hydrogen sulfide that was being produced because there was no oxygen in the river. Hydrogen sulfide can only be produced in the absence of oxygen. Another consequence was that even the most pollution-tolerant fish like carp could not survive.

Concerned residents of Melvindale and Dearborn got together and started a petition drive in 1984, and almost 1,000 signatures were gathered. Residents demanded that the U.S. Environmental Protection Agency and the Michigan Department of Natural Resources do something to fix this problem in the Rouge River.

It is interesting that from this grassroots petition drive the Rouge River would go from having a reputation of being one of the most polluted rivers in the United States to gaining a national reputation for its efforts in watershed management. This story is important not only because of how the Rouge River impacts the Detroit River (the Rouge River is a tributary of the Detroit River), but because these watershed management efforts need to be duplicated in other watersheds. Watershed management efforts represent the necessary building blocks to restore and maintain the Detroit River and the Great Lakes.

Early History

From the very beginning the Rouge River has been recognized for its strategic location and beneficial uses. Native people settled along the shores of the river to use it as a transportation route, take advantage of the water supply, and benefit from the incredible fishing and hunting. Several tribes settled in the area and lived in harmony with their environment.

When the French arrived in the late seventeenth century they named the Rouge River for its natural red clay color. The voyageurs, couriers du bois, explorers, and trappers all used the Rouge River as a passage from the Detroit River to the interior. In 1701 the French built their fort and established the first permanent settlement in the region.

During the 1700s the Rouge River was a favorite stomping ground of French pony owners. They raced their ponies on the frozen riverbed from the "first bend" near where the current Henry Ford estate stands to "World's End" where Fort Street now crosses the Rouge River. What followed was the development of land in the form of strip farms along both the Detroit and Rouge Rivers. Later in the eighteenth century gristmills would be established on main forks in the river and shipyards established on the lower river. In fact, twenty vessels were launched on the Rouge River between 1770 and 1780. The Rouge's contribution to shipbuilding continued through World War I, when eagle boats were built to fight submarines. These two-hundred-foot-long submarine chasers were the first ships ever made by mass production.

The lower Rouge River was dredged in the early 1900s to accommodate the building of the eagle boats and the needs of automobile manufacturing. Henry Ford consolidated automobile manufacturing operations into one geographical area to create efficiencies and synergies. The four-hundred-acre Zug Island became the heartland of the Industrial Revolution.

When industrial operations became established, the Rouge River became the recipient of waste oil and other industrial pollutants. By the 1960s the river was flowing orange with industrial pickle liquor waste. However, that orange color was often only evident when a boat cut a wake through the heavy waste oil floating on the surface (Cowles 1975). During this time the Rouge River became infamous as one of three Great Lakes tributaries to catch on fire. The other two were the Cuyahoga River in Cleveland and the Buffalo River in New York.

As a result of the very visible oil pollution, Michigan initiated an industrial pollution control program in the 1960s. Later Michigan would partner with the federal government to establish a National Pollutant Discharge Elimination System Program. The result of these early industrial pollution control efforts was that the Rouge was less polluted with oil and appeared somewhat cleaner. But as we have all learned, it is what you cannot see that you need to worry about.

In a classic study of the Rouge River in the early 1970s, the Michigan Department of Natural Resources (Jackson 1975) reported that "approximately 40 miles of the Rouge River were characterized by very poor water quality as evidenced by a macroinvertebrate community dominated by animals tolerant of severely polluted waters. The principal contaminants at that time were raw sewage and inorganic sediment entering the river via combined and/or storm sewers."

Much of this river was grossly polluted and did not meet state water quality standards for municipal discharges, which only became evident after the initial effort to control industrial discharges. Not much changed by the early 1980s. The Rouge River was still highly polluted—raw sewage was still being discharged, there were odor problems, and fish were dying. In addition, the Rouge River was negatively impacting the Detroit River. For example, immediately downstream of the Rouge River, pollution-tolerant worms numbered over one million per square meter of river bottom in both 1968 and 1980, demonstrating long-term, severe organic enrichment (Thornley and Hamdy 1983).

The next important event in the history of the Rouge River was the 1984 petition from residents in Melvindale and Dearborn. As a result, the Michigan Department of Natural Resources and the Michigan Water Resources Commission commissioned a staff report of the status of water quality and pollution control efforts in the Rouge River Basin. That Michigan Department of Natural Resources study reported that "the lower Rouge system is plagued with natural and manmade factors which limit water quality to the extent that approximately 40 river miles do not meet Michigan water quality standards" and led the Michigan Water Resources Commission to request the development of a Rouge River Basin Strategy, an action plan, and a public participation process to abate water pollution problems in the entire Rouge River Basin (Hartig 1984). The Rouge River Basin Strategy, action plan, and public participation process were adopted on October 1, 1985. All forty-eight communities agreed to participate in a watershed-based process to develop and implement a comprehensive remedial action plan (RAP) to restore all impaired beneficial uses over a twenty-year period.

A significant related event also occurred in 1985. In a tragic accident, a twenty-three-year-old man fell into the polluted Rouge River, swallowed water, and died of an infection from a rare parasitic, waterborne disease (Diebolt 1985). The man died of leptospirosis or rat fever. Even though a local health department news release stated that there was no evidence that the man's illness and death were directly related to water quality in the Rouge River, most people knew that the probable route of exposure was the river. Governments have always said "show us the human health impact." There was now a "compelling weight of evidence" that the Rouge River was the probable source of the waterborne pathogen that caused one man's death. The health department had to warn the public to avoid contact with the river. The stakes had risen.

Watershed Management

The Rouge River RAP process began in 1985 with the recognition that state government had the clout to force corrective actions. However, this enforcement approach could result in lengthy litigation and delayed cleanup and not in the local ownership of the RAP necessary for implementation. The Rouge River Basin Committee was established to work with local governments and stakeholder groups and bring about local ownership of the RAP. All 48 communities in the 467–square mile (1,210 km^2) watershed participated on the Rouge River Basin Committee along with other stakeholders like the League of Women Voters, industries, and environmental groups. An Executive Steering Committee was also established at this time to direct and oversee the RAP process.

The initial Rouge River RAP was completed in 1988 and represented a blueprint for restoring all impaired beneficial uses (SEMCOG 1988). It is no surprise that this initial RAP focused heavily on sanitary sewer improvements and combined sewer overflows because approximately 7.8 billion gallons of combined sewage were being discharged into the Rouge River annually from 168 combined sewer overflow points.

In 1992 the Rouge River National Wet Weather Demonstration Project (Rouge Project) was established with funding from the U.S. Environmental Protection Agency to Wayne County's Department of Environment to implement the RAP (Murray 1994). The Rouge Project recognizes the importance of addressing wet weather pollution problems in the river and developing a cost-effective watershed approach to deal with them. This comprehensive implementation program deals with the problems of combined sewer overflows, polluted storm water runoff, and various other nonpoint source pollutants. Also included in the Rouge Project is an intensive monitoring and sampling program, modeling, geographic information systems, and public outreach, all of which build the capacity for watershed management. The Rouge Project also serves as a model for other watersheds throughout the country.

In 1994 the Rouge River RAP was updated to catalog progress, identify additional remedial and preventive actions necessary to restore uses based on what had been learned through the Rouge Project and elsewhere, and to fully adopt an ecosystem approach to solving all problems (Michigan Department of Natural Resources and SEMCOG 1994). An ecosystem approach accounts for the interrelationships among water, land, air, and all living things, including humans, and involves all user groups in management. The initial institutional structure called the Rouge River Basin Committee was replaced with a Rouge RAP Advisory Council (RRAC) and a series of subcommittees to address more specifically issues and causes of use impairments. The RAP was updated again in 1998 in the spirit of continuous improvement (Michigan Department of Environmental Quality 1998).

Public participation, education, and outreach have been championed by a nonprofit organization called Friends of the Rouge. Since 1986 their mission has been to get people directly involved in cleaning up and restoring their river. They have worked tirelessly to educate both young people and adults about the causes and effects of river pollution, promote understanding of what can be done about it, and achieve a sense of "ownership." Examples of grassroots projects championed by Friends of the Rouge include:

- an annual stream cleanup extravaganza called Rouge Rescue
- a comprehensive RiverWatch program for helping volunteer

groups "adopt" a local stream and take charge of reducing pollution in their neighborhoods

- a stream monitoring program that began in 1987 with 16 Detroit-area high schools and has expanded to over 75 elementary, middle, and high schools
- a storm drain stenciling program that informs local residents that any pollutants (such as lawn chemicals or motor oil) that enter storm drains go straight into their streams without treatment
- workshops on subjects such as building nesting boxes for birds, ducks, and bats
- a frog and toad survey program designed to identify and monitor wetlands and other valuable wildlife habitats
- a River Steward program to attract and educate citizens for restoration and stewardship efforts
- habitat enhancement projects such as soft engineering of shorelines
- a program of disconnecting downspouts from storm sewers, installing a rain barrel, or diverting rainwater away from paved surfaces to reduce the "flash-flood" effect that contributes so much to pollution and erosion problems

Making It Happen and Achieving Results

Considerable progress has been made as a result of the Rouge River RAP, the Rouge Project, and other efforts. All of the sanitary sewer capacity improvements and nearly all of the Phase 1 combined sewer overflow (CSO) projects called for in the 1988 RAP have been completed. Since 1988, over $900 million has been spent on sewer capacity improvements and combined sewer overflow projects. Congressman John D. Dingell has been instrumental in helping secure federal funding to initiate and sustain efforts under the Rouge River National Wet Weather Demonstration Project and the RAP.

Sewer separation projects have been completed and 9 CSO retention/treatment basins are now operational, eliminating and/or controlling 76 of the 157 CSO outfalls in the Rouge River watershed. The process for evaluating the CSO basin performance is underway. Based on evaluation results thus far, the Michigan Department of Environmental Quality has concluded that the 6 CSO treatment facilities for which data are available are meeting the evaluation team's criteria for the elimination of raw sewage and the protection of public health. This conclusion will have significant impacts on the design and construction of future CSO treatment facilities in the Rouge watershed, as well as in Michigan and the rest of the country. As a result of the CSO control program to date, 130 miles of stream are now free of uncontrolled CSO discharges, and the completed basins are controlling overflows at a rate of approximately 4 billion gallons per year.

As the pollution caused by sanitary and combined sewer overflows has been substantially reduced, other sources of pollution and resource degradation (e.g., urban storm-water runoff, illicit connections, failing septic systems, low flow conditions, habitat loss and degradation) have become more apparent and important. It is generally accepted that implementation of nonpoint source (i.e., diffuse sources of pollution, such as runoff from precipitation, that do not enter a river at a fixed point) control practices and environmentally sustainable land use practices are best handled at the local level. Therefore, considerable efforts have been made to establish a regulatory framework to address these other sources of pollution concurrent with future sanitary sewer and combined sewer overflow control efforts.

With the support of the federal court and through the leadership of the Wayne County Department of Environment, the Rouge Program Office and the local communities assisted the state of Michigan with the development of the Michigan Voluntary General Storm-water Permit. These voluntary storm-water permits are the first issued in the nation to control stormwater runoff. Forty-three local units of government, counties, and agencies, representing over 95 percent of the Rouge River watershed land area, have subsequently applied for permit coverage. All permittees have established public education and illicit discharge elimination programs as required by the permit. Subwatershed advisory groups have now been formed under the general permit process to address local issues relating to sanitary and combined sewer overflows, polluted storm-water runoff, flow management, habitat, and other locally identified issues. These subwatershed advisory groups have developed subwatershed plans, which will form the foundation of the next revision to the RAP. These plans are now being implemented. Success in this next phase of community-based, subwatershed management will in large part be dependent on the success of these subwatershed advisory groups and the voluntary storm-water permits.

Through Rouge Project funding, a total of 120 pilot storm water management projects are being implemented or have been implemented throughout the watershed by over 30 different communities and agencies. Types of projects underway include:

- creation and restoration of wetlands
- rehabilitation of upland areas into grassed swales and detention ponds

- elimination of illicit discharge connections
- incorporation of best management practices to minimize erosion controls
- stabilization of stream banks
- rehabilitation and enhancement of habitats
- public information and involvement projects

Technical support will be critical to the success of the subwatershed advisory groups and the general permit. The Rouge Program Office, working with the counties and the U.S. Geological Survey (USGS), has initiated a long-term watershed monitoring network including rain gauges and continuous in-stream water quality and flow measurement stations. Volunteer monitoring efforts are being integrated into the long-term program. Computer modeling and a comprehensive geographic information system (GIS) have been established to assist with watershed management decision making and assessment. Monitoring that assesses the current state of the river and its watershed is considered a critical element because it:

- establishes baseline conditions
- supports the development and use of watershed models
- identifies problems and their sources
- evaluates effectiveness of control programs
- helps measure progress and celebrate successes

Through the long-term monitoring program the success of current CSO and general permit watershed management activities is being identified. Improvements are being documented in dissolved oxygen levels throughout the watershed. For example, during 1999 there were no violations of the 5 mg/L dissolved oxygen standard during dry weather conditions at the Greenfield Road monitoring station in the concrete channel. In 2000, the dissolved oxygen standard was met 94 percent (or more) of the time in both wet and dry periods at monitoring stations along each branch of the Rouge River. Such monitoring results are supported by reports of an improved fishery in the Rouge River.

The Future: Promoting a Vision and Changing Perceptions

It is amazing that the Rouge River has gone from being known as one of the most polluted and poorly managed rivers in the United States to being a model for community-based watershed management. However, although much has been accomplished, much remains to be done. Heightened pressure from ever-increasing urbanization can destroy habitat and decrease fish, wildlife, and other aquatic populations. More effort needs to be expended on preserving critical habitats and ensuring that new developments are environmentally sustainable. Controlling storm water and minimizing its impacts will be a priority for some time. It will be difficult because sources are widespread, diverse, and numerous. Low stream flow is a periodic problem in the Rouge River, which can have substantial negative impacts on fish and other aquatic life. Historic U.S. Geological Survey flow data show an increasing base flow trend even in light of development. Rapid high flow conditions also exist because development pressures are increasing the percent of impervious surfaces, which in turn creates more runoff. This factor has been cited as one of the major causes of decreased fish and aquatic invertebrate populations in the river.

A number of recent developments have occurred that are changing the perception of the Rouge River and promoting a positive vision. Examples include:

- Water-based recreation (Newburgh Lake restoration, triathalon, Rouge River canoeing): Because of the initial success of sanitary sewer capacity improvements, CSO sewer separation projects, and documented water quality improvements, Wayne County Parks opened a canoe livery in 1996 downstream of Newburgh Lake in the Middle Rouge River (figure 13.1). This was the first time in over twenty-five years that body contact recreation was encouraged on the river, clearly manifesting that the river was making a comeback. Since that time, Newburgh Lake has been dredged and restocked with game fish at a cost of approximately $12 million. In 2000, Wayne County hosted the first annual Newburgh Lake Triathalon. Canton Township, along the Lower Rouge River, has also indicated interest in opening another Wayne County Parks canoe livery.

- Johnson Creek protection and designation as a cold-water stream: This stream is in the headwaters of the Middle Rouge River and was recently recognized as one of only three streams in southeast Michigan capable of supporting trout. The Michigan Department of Natural Resources has been stocking the stream with brown trout since 1994 and recently changed its status from a warm-water to a cold-water stream. A very active public/private partnership has been established in the form of the Johnson Creek Protection Group, whose mission is to "preserve, protect, restore, and

Figure 13.1. A canoe livery recently opened on the Rouge River. (Wayne County Rouge River National Wet Weather Demonstration Project)

enhance the water quality, habitat, and function of this cold-water stream and its watershed."

• Fishery targets, recovery of the fishery, and return of stoneflies: With funding from the Rouge Project, the Michigan Department of Natural Resources-Fisheries Division and the Institute of Fisheries Research at the University of Michigan have completed an extensive fisheries study of the Rouge River. Results of the research provide important fishery targets that the subwatershed advisory groups have integrated into their subwatershed management plans and monitoring goals. Survey data on fish and other aquatic organisms throughout the watershed indicate that further recovery of the fishery is indeed achievable. Salmon and steelhead trout are being found in upstream segments of the Lower Rouge River (figure 13.2). Stoneflies have recently been found by Friends of the Rouge volunteers in both Johnson and Fowlers Creeks. Stoneflies are indicator organisms that reflect good water quality.

Figure 13.2. A steelhead trout caught in the Rouge River. (Wayne County Rouge River National Wet Weather Demonstration Project)

- Rouge Gateway partnership: One of the most ambitious developments in the efforts to further restore the Rouge River is a community-business partnership called the Rouge Gateway Project. This project will undertake ecological restoration in a seven-mile section of the Lower Rouge River from Michigan Avenue in Dearborn to the Detroit River (figure 13.3). This reach represents the most degraded portion of the river and includes a four-mile concrete channel and a three-mile section of navigable dredged waterway downstream of the channelized section

The Rouge Gateway Project will have three phases:

Phase 1: Planning. This phase will involve all stakeholders in establishing guidelines for the restoration of wetlands, riparian shoreline, and fish and wildlife habitat consistent with goals of the Michigan Department of Natural Resources. Public uses along the river will be expanded, including walkways, parks, and water tours.

Phase 2: Early Restoration Efforts. One or two areas along the concrete channel and within the public-right-of-way will be

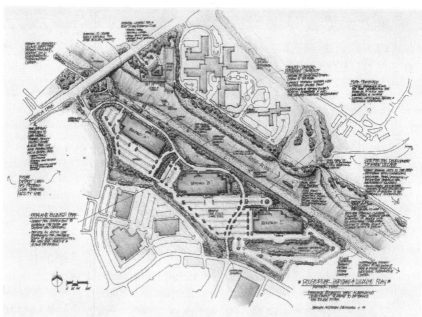

Figure 13.3. The concrete channel of the Lower Rouge River as it exists today (top) and a graphic depiction of what this portion of the lower Rouge River might look like in the future (bottom). The vision for this area includes greenways, parks, soft engineering of the shoreline, and mixed-use redevelopment. (Wayne County Rouge River National Wet Weather Demonstration Project)

selected as pilot projects for demonstrating use of soft engineering techniques (i.e., using ecological principles and practices to achieve stabilization of shorelines and public safety, while enhancing habitat, improving aesthetics, and saving money).

Phase 3: Full Restoration. The final phase will continue the restoration work along the concrete channel and the navigable part of the river downstream to the Detroit River.

Phase 1 planning began in 2000. The tentative target date for completion of the early restoration efforts is 2003. The Wayne County Department of Environment is working with the U.S. Army Corps of Engineers on feasibility, design, and implementation under phases 2 and 3. A tentative target date for completion of the full restoration work is 2010.

Concluding Remarks

The Rouge River has become a national model for watershed management. However, continued technical and management support must be provided to the subwatershed advisory groups so that people are knowledgeable about their river's problems and engaged in solving them. Friends of the Rouge must continue to play a key role in education and outreach. It will only be through informed and engaged stakeholders that we will be able to care for our rivers as our homes and reap their potential economic, societal, and recreational benefits.

REFERENCES

Cowles, G. 1975. "Return of the River." *Michigan Natural Resources* 44.1:2–6.
Diebolt, J. 1985. "Bad Rouge Water May Have Killed Novi Man." *Detroit Free Press,* October 5.
Hartig, J. H. 1984. *Status of Water Quality and Pollution Control Efforts in the Rouge River Basin.* Lansing: Michigan Department of Natural Resources.
Jackson, G. 1975. *A Biological Investigation of the Rouge River, Wayne and Oakland Counties, May 17–October 19, 1973.* Lansing: Michigan Department of Natural Resources.
Michigan Department of Environmental Quality. 1998. *Rouge River Remedial Action Plan Progress Report.* Livonia, MI.
Michigan Department of Natural Resources and Southeast Michigan Council of Governments (SEMCOG). 1994. *Rouge River Remedial Action Plan Update.* Detroit.
Murray, J. 1994. *Rouge River Watershed Management: Implementing a Remedial Action*

Plan (RAP). Proceedings of 67th Annual Conference of Water Environment Federation. Alexandria, VA.

Rouge Remedial Action Plan Advisory Council. 1999. *Rouge River Report Card*. Detroit: Wayne County Department of Environment.

SEMCOG. 1988. *Remedial Action Plan for the Rouge River Basin*. Vol. 1: *Executive Summary*. Detroit.

Thornley, S., and Y. Hamdy. 1983. *An Assessment of Bottom Fauna and Sediments of the Detroit River*. Toronto: Ontario Ministry of Environment.

Wayne County Rouge River National Wet Weather Demonstration Project. 2001. *Review of Year 2000*. Detroit: Wayne County Department of Environment.

THE GREATER DETROIT AMERICAN HERITAGE RIVER INITIATIVE AND THE FUTURE

John H. Hartig

The Detroit River has faced a number of environmental and natural resource problems, including:

- exploitation of beavers (chapter 4)
- waterborne disease epidemics (chapter 5)
- oil pollution (chapter 6)
- phosphorus pollution (chapter 7)
- toxic substances contamination (chapter 8)
- loss of habitat and biodiversity (chapters 9, 10, and 11)

The problems have changed along with human uses and abuses of the Detroit River ecosystem and with the evolution of management strategies. These strategies have evolved from narrow, single-purpose objectives (e.g., eliminating waterborne disease epidemics, stopping oil pollution) to more comprehensive, multipurpose objectives (e.g., restoring all beneficial uses, including drinking water, swimming, industrial water supply, fish and wildlife health, habitat, etc.). Further, management strategies have evolved from top-down, command-and-control programs led by federal and state/provincial governments to bottom-up, multi-stakeholder partnerships.

From a national perspective, there have been three distinct eras that have had a lasting impact on conserving natural resources and protecting the environment. The nation has benefitted from each one, and each new era was built on the practical foundation of previous ones.

Most people recognize President Theodore Roosevelt's significant contribution to conservation of special places during his presidency (1901–1909). He was a man who acted on his convictions with vigor. When business interests threatened America's natural resources, Roosevelt moved to protect them using executive powers. He set a precedent of using the Antiquities Act of 1906 to create scores of national monuments, refuges, and parks, including Grand Canyon National Monument, the Tongass forest reserve, and Muir Woods. He championed a balanced approach toward preservation and sustained productivity of natural resources like forests and rivers. He was a strong, progressive leader who promoted recreation in the wilderness to enhance personal growth. There is no doubt that Roosevelt's enduring legacy is an expanded national conservation system for national monuments, refuges, and parks.

Another important era was catalyzed when Rachel Carson published *Silent Spring* in 1962. She was a marine biologist with the U.S. Bureau of Fisheries and later became editor of all publications for the U.S. Fish and Wildlife Service. Her passion was writing. *Silent Spring* is a scholarly and well-written book documenting how pesticides like DDT and other chemicals that were used to enhance agricultural productivity were also poisoning our rivers, lakes, oceans, and all forms of life. This landmark book started the environmental movement and eventually led to Earth Day. *Silent Spring* would become part of the vocabulary of the nation. This environmental era began with the banning of certain hard pesticides and eventually led to the establishment of strong federal and state programs that would control chemical pollutants. Indeed, the 90 percent decline in DDT contamination of Great Lakes fish can be directly traced to the environmental movement catalyzed by Rachel Carson's *Silent Spring*.

We have now entered a new era that emphasizes sustainability and healthy communities. Historically, many federal and state programs were dictated to communities in a top-down, command-and-control fashion. For example, thirty years ago command-and-control, regulatory, pollution control programs made both environmental and economic sense because of higher pollution levels. However, once the minimum technological requirements were established for specific industries and municipalities, this strategy produced a reduced rate of success and higher costs as more complex and expensive controls were adopted to capture or control smaller amounts of pollution. The transactional costs and adversarial relationships associated with administering, permitting, and enforcing such a system rose substantially over time.

Today, more and more programs are being implemented in a bottom-up, cooperative fashion. Further, these new programs are striving for a balanced approach among environment, economy, and community. Federal and state governments are providing new tools and resources, cutting red tape, and helping deliver community priorities. Communities are now identifying the priorities, and federal and state/provincial governments are working in a value-added mode to help deliver the community priorities. These priorities typically address quality of life and environmentally sustainable economic growth.

Community-Based Programs

Community-based programs are designed to achieve sustainable and healthy communities. They are aimed at helping communities grow in ways that improve quality of life and achieve strong, sustainable economic growth. The role of government has changed from one of top-down leadership to value-added partnership. The intent is to expand choices available to communities, not to dictate solutions. By providing new tools and resources, federal and state governments can help communities create the future they want.

Community-based programs share a number of common principles (U.S. EPA 1999) including:

- focusing on a definable geographic area
- working collaboratively with a full range of stakeholders through effective partnerships
- assessing the quality of air, land, water, and living resources in a place as a whole
- integrating economic, environmental, and social objectives and fostering local stewardship of all community resources
- using the appropriate public and private, regulatory and nonregulatory tools
- monitoring and redirecting efforts through adaptive management

Frequently such community-based programs have a number of strategic goals:

- preserving green spaces that promote clean air and clean water, sustain wildlife, and provide families with places to walk, play, and relax
- easing traffic congestion by improving road planning, strengthening existing transportation systems, and expanding the use of alternative transportation

- restoring a sense of community by fostering citizen and private sector involvement in local planning
- promoting collaboration among neighboring communities— cities, suburbs, or rural areas—to develop regional growth strategies and address common issues like crime
- enhancing economic competitiveness by nurturing a high quality of life that attracts well-trained workers and cutting-edge industries

One good example of a community-based program is the American Heritage River Initiative. In 1998, fourteen American Heritage Rivers were designated by Executive Order from 126 community nominations. But why give a special designation to rivers?

Charles Kuralt was a noted writer, reporter, anchor, editor, and storyteller. He was best known for exploring America, talking to people, and sharing their stories. His over six hundred episodes of "On the Road" gave him a unique perspective on American places and people. He shared his perspective on the importance of rivers in our United States: "I started out thinking of America as highways and state lines. As I got to know it better, I began to think of it as rivers. Most of what I love about the country is a gift of the rivers. . . . America is a great story, and there is a river on every page of it." All life needs water, and rivers are the water supply of the earth.

The Greater Detroit American Heritage River Initiative

The Detroit River was honored to become one of the fourteen American Heritage Rivers. Specifically, this program helps communities restore and revitalize waters and waterfronts. The heart of the American Heritage River Initiative is locally driven and designed solutions. The federal government role is fostering community empowerment and helping provide focused attention and resources to help river communities revitalize their economies, renew their culture and history, and restore their environment. The initiative integrates the economic, environmental, and historic preservation programs and services of federal agencies to benefit communities engaged in efforts to protect and enhance their rivers. Further, it encourages investment in river communities, promoting partnerships and leveraging of state, nonprofit, and business resources.

In this new millennium, the Detroit area is poised for significant and sustainable renewal. Sustainability is now a large part of the context of this rebirth, as noted by the fact that Detroit hosted the National Town Meeting for a Sustainable America in May 1999 (Global Environment & Technology Foundation 1999).

Sustainable development is a major goal of both the public and private sectors. It calls for meeting the needs of the present generation without compromising the ability of future generations to meet their needs. At a practical, working level, it is viewed as balanced and continuous economic, social, and environmental progress. The Greater Detroit American Heritage River Initiative is a multi-stakeholder process designed to achieve, sustain, and celebrate our communities, economies, histories, cultures, and environments. It is viewed as practical application of sustainable development at the community level.

Oversight of the initiative is provided by an executive committee composed of representatives of: the city of Detroit, twenty-one downriver communities, Wayne County, and the business community. A multi-stakeholder steering committee and business industry group are also in place to provide advice to the executive committee. Project management and administrative support is provided by a public-private partnership of business, labor, and governmental leaders called the Metropolitan Affairs Coalition.

The Greater Detroit American Heritage River Initiative strives for continuous improvement. Figure 14.1 presents a model of the implementation framework being used by the initiative. This framework is guided by nine important principles:

- stakeholder involvement
- commitment of top leaders
- agreement on information needs
- strategic planning
- implementation
- human resource development
- monitoring results and celebrating milestones
- review and evaluation
- stakeholder satisfaction

Consistent with the framework in figure 14.1, stakeholders were initially brought together to develop the American Heritage River (AHR) application. As part of the application development, stakeholders agreed to the following community vision:

> We are filled with pride for our magnificent River and have a shared vision for its regeneration. Our vibrant international waterway inspires a community brimming with fun and excitement, rest, and relaxation. A broad diversity of jobs,

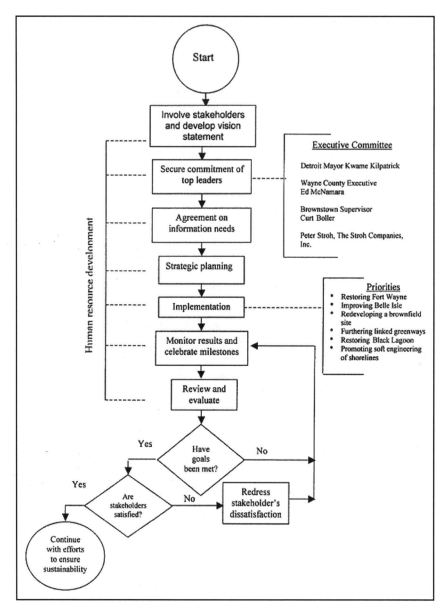

Figure 14.1. A project implementation framework being used by the Greater Detroit AHR Initiative.

housing, historic interpretation, recreation, and culture breathe life into a 24 hour a day waterfront. Industry, commerce, and tourism growing in harmony with the environment sustain fish, bird, animal, and plant habitats. The River has become the region's front door, with access to all inland communities. Its beauty and integrity have been restored, and we thrive within its ecosystem.

Commitment was then sought from top leaders. Once commitments were obtained, agreement had to be reached on information needs to begin strategic planning. The executive and steering committees reviewed over 150 potential projects to be able to establish six first-phase priorities. Both short- and long-term projects are envisioned to help ensure progress and build a record of success. In this framework, priorities are set, projects implemented, and results measured in an iterative fashion for continuous improvement. Planning and implementation can proceed simultaneously; actions can be taken before all plans are fully complete. Results are evaluated against milestones and benchmarks to measure and sustain progress. Improvements in the process can be made to ensure that the desired outcomes are achieved within appropriate time lines. Frequent and rigorous review and feedback are necessary to ensure that the process stays on track; mid-course corrections are made where necessary. Stakeholder satisfaction is also measured in this process.

As shown in figure 14.1, human resource development is integrated throughout the process to reinforce the need for cooperative learning among all stakeholders. It is believed that such a process, if followed, will lead to achievement of the Greater Detroit AHR Initiative goals and community vision. As noted above, the executive committee, upon the advice of the steering committee, approved six first-phase priorities.

Improving Belle Isle

Belle Isle is a 980–acre island park located in the Detroit River within close proximity to the central business district of the city of Detroit (figure 14.2). It is the crown jewel of Detroit's public park system. Belle Isle was designed in 1883 by Frederick Law Olmstead, who also designed New York's Central Park. It is situated on one of America's busiest waterways and provides spectacular views of Detroit, Canada, freighter traffic, and the Ambassador Bridge. Other unique features of Belle Isle include:

- a 200–acre forest that is one of the highest quality native woodland environments in southeast Michigan

Figure 14.2. An aerial photograph of Belle Isle (U.S. Army Corps of Engineers).

- the Belle Isle Aquarium, the oldest aquarium in America
- the Nancy Brown Peace Tower
- the Dossin Great Lakes Museum, which has the world's largest collection of scale-model Great Lakes ships
- the Belle Isle Nature Center
- the Scott Memorial Fountain
- the Anna Scripps Whitcomb Conservancy
- the Belle Isle Zoo

Belle Isle has been a free island park for over 150 years. During the past three decades, the declining tax base in the city has led to reduced funds available for park maintenance and capital investments. The park is also heavily used—eight million annual visitors—making it easily the busiest park in the region and in the state. The combination of insufficient capital and overuse has taken its toll on Belle Isle; many of its facilities have fallen into a serious state of disrepair.

The Detroit Recreation Department has made substantial efforts to plan for the necessary improvements to the island through a multiyear master planning process. This process has identified the $180 million in capital investment needed to prepare the island to fulfill its role as Detroit's premier open space. One potential source of revenue for this program has been proposed as an auto entry fee of $3/car or $20/year. A Belle Isle toll would generate substantial funds for improving the park consistent with the master plan and would also leverage substantial federal and state funds in the process. This toll option has been the subject of heated debate in the city. Other sources of funds include state and federal agencies with recreation, conservation, and environmental mandates. The toll option would not only provide much needed operation and maintenance dollars, but would help the Detroit Recreation Department make the match requirements on federal and state grants.

Consistent with an adaptive management process where priorities are set, actions taken, and results measured in a continuous improvement fashion, the Detroit Recreation Department and the Greater Detroit AHR Initiative have been working the U.S. Army Corps of Engineers on stabilizing shorelines around the island. In addition, they will be working with other public and private partners to reconfigure a road, improve trails, and enhance wildlife habitats consistent with the master plan. In 2002, Detroit Recreation Department received a $500,000 Urban Park and Recreation Recovery Program grant from the National Park Service to restore Flynn Pavilion and reopen it as a canoe-bicycle-kayak-ice skating facility that has historically been part of the Belle Isle experience.

However, a stewardship ethic is also required. The Detroit Recreation Department and the Greater Detroit AHR Initiative have been sponsoring a series of hands-on projects like restoring habitat at Blue Heron Lagoon and rehabilitating the shoreline at Lake Muskoday. Each of these projects is aimed at developing a stewardship ethic for Belle Isle. Belle Isle Park Superintendent Marchel McGehee views the development of this ethic as an essential ingredient in helping implement the proposed Belle Isle master plan: "The Detroit Recreation Department needs a committed core of volunteer stewards to help keep Belle Isle clean and green for future generations. There is no better way of developing volunteer stewards than to work with our children." (Press release, Sept. 18, 2000)

Restoring Fort Wayne

Historic Fort Wayne was built in the early nineteenth century during a period of tension with the British in North America. It was used during the Civil War and both world wars. Later, it became one of the largest induction centers during the Korean and Vietnam Wars.

Strategically located at a bend in the Detroit River, it comprises over 83 acres and has over 30 buildings. It is also adjacent to an empowerment zone. The city of Detroit, Wayne County, and the Huron Clinton Metropolitan Authority have established a nonprofit corporation called the Historic Fort Wayne Park and Museum Council to cooperatively restore, operate, and maintain the fort. This was a major breakthrough in sharing responsibility for restoration, operation, and maintenance. Other business and community partners will be brought on in the future to build a larger financial base to take care of this historic jewel.

Fort Wayne was closed from 1992 to 2000 and reopened in 2001 on a limited basis. On July 22, 2001, Historic Fort Wayne was opened for 18,500 people to watch the tall ships parade held as part of the three hundredth anniversary of European settlement of Detroit.

Historic Fort Wayne is being managed for historical interpretation, education, recreation, and community activities. The nonprofit corporation partners have reached agreement on these desired uses and a cooperative management structure. The reopening and restoration of Fort Wayne represents a significant step toward revitalization of southwest Detroit. Further, it helps celebrate an important part of Detroit's history and opens up approximately one mile of riverfront. Economic development will also be enhanced with a transient marina and waterfront restaurant. Three grants received for Historic Fort Wayne include:

- a $46,000 Coastal Zone Management grant (2001) to evaluate the shoreline, including restoration of a portion to its natural state
- a $98,000 Community Foundation of Southeast Michigan-Greenway Initiative grant (2001) to design greenway trails between the fort and adjacent community parks
- a $48,000 Michigan State Historic Preservation Office grant (2002) to repair the roof of the theater building at the fort

Redeveloping a Brownfield Site

Brownfields are vacant industrial parcels of land that are usually contaminated and provide little or no tax revenue or other benefits to a community. Like all major urban areas, southeast Michigan has many brownfields that have enormous potential for redevelopment. The Greater Detroit AHR Initiative is working on a brownfield redevelopment project that would serve as a model for the region.

The proposed brownfield site is the former McLouth Steel Company property in the cities of Riverview and Trenton. It is nearly 300 acres in size, has over 6,100 feet of river frontage, and is located along Jefferson Avenue, the region's riverside drive. For the last several years, a small part of this once giant steel-making site has been used by a company called DSC.

The city of Trenton has been working with McDonough and Partners (chapter 12) to develop a conceptual plan for mixed-use redevelopment consistent with the transition along modern riverfronts to non-industrial uses. The Greater Detroit AHR Initiative is working in partnership with federal, state, and county agencies to help further mixed-use redevelopment of this site consistent with community goals. Four federal agencies have agreed to be a partner in a model brownfield redevelopment project at this site. Federal and state resources can be leveraged for site evaluation, technical assistance for planning, and remediation. Although an enormous task, the brownfield redevelopment of the DSC site offers one of the premier opportunities along the Detroit River to reclaim and reuse a former industrial site. Initial efforts are focused on acquiring the northern 75 acres for commercial and recreational development. The inclusion of greenway trails and some parkland would provide Riverview with its first waterfront park. Some commercial development would provide tax revenues to the cities.

Restoring Black Lagoon

Black Lagoon is located along the Trenton Channel of the Detroit River in Trenton, Michigan. Contaminated sediment in the Black Lagoon has some of the highest acute toxicity in the river and has therefore been identified as a priority site for remediation. Sediment in the Black Lagoon is contaminated with oil, grease, and heavy metals, including high levels of mercury. Approximately $1 million has been allocated from Michigan's Clean Michigan Initiative and the U.S. Environmental Protection Agency's Great Lakes National Program Office for removal and disposal of 30,000 cubic yards of contaminated sediment from Black Lagoon. Approximately 5,000 cubic yards will be treated by ENDESCO (as a wholly owned subsidiary of the not-for-profit Institute of Gas Technology) using their patented Cement-Lock process. This technology will destroy organic contaminants (PCBs, oils, and grease) and immobilize the remaining contaminants (heavy metals) for reuse as construction-grade cement.

The cost of demonstrating the Cement-Lock process on 5,000 cubic yards of contaminated sediment is approximately $9 million, of which $3 million will likely come from state and federal funding sources. The remaining 25,000 cubic yards will be disposed in a confined disposal facility or an appropriate upland disposal facility.

Under the Greater Detroit AHR Initiative, Trenton and the U.S. Army Corps of Engineers-Detroit District have investigated under section 206 of the U.S. Water Resources Development Act alternatives for restoring its riverbank and nearshore aquatic habitat at Black Lagoon. Technical support was provided by the U.S. Geological Survey-Great Lakes Science Center, Michigan Department of Natural Resources, U.S. Fish and Wildlife Service, Michigan Department of Environmental Quality, and U.S. Environmental Protection Agency.

This project was unique in that it used the opportunity of sediment remediation to achieve habitat restoration and shoreline improvements. It is an excellent example of interagency cooperation in use of an ecosystem approach to simultaneously achieve goals for sediment remediation and habitat restoration. The result will not only be a cleaner lagoon but an improved shoreline and enhanced habitat. The preferred option is to re-create an island that was historically lost to expansion of a steel plant. This all adds up to creating a unique riverfront destination that improves the quality of life for residents. Upon completion of this project, the community wants to hold a contest to rename Black Lagoon to further celebrate its revitalization.

Promoting Soft Engineering of Shorelines

Historically, many river shorelines were stabilized and hardened with concrete and steel to protect developments from flooding, stop erosion, or accommodate commercial navigation or industry. Typically shorelines were developed for a single purpose. Today there is growing support for development of shorelines for multiple purposes.

Up and down our Detroit River, efforts are underway to reshape the riverfront from something concealed in our backyard to the focal point of our attention. General Motors has switched the front door of the Renaissance Center from Jefferson Avenue to the Detroit River with the building of a five-story Wintergarden. The promenade stretching east from the Renaissance Center further showcases our river for businesses and residents. In Windsor another three miles of continuous riverfront greenway were opened in 1999 to promote our river and help create an exciting venue for people to work, play, and socialize downtown. In Wyandotte the building of a golf course, a rowing club, and greenways have directed attention to our river and have resulted in considerable spin-off benefits. People want to increase access to our river, incorporate trails and walkways to it, improve the aesthetic appearance of the shoreline, and reap recreational, ecological, and economic benefits from it.

One of the first things that was done under the Greater Detroit AHR Initiative was to convene a major conference to consider how to reshape the Detroit River shoreline using techniques of soft engineering. Hard engineering of shorelines is generally defined as use of concrete breakwalls or steel sheet piling to stabilize shorelines and achieve safety (Caulk et al. 2000). There are many places along our working river where hard engineering is required for navigational or industrial purposes. As most people know, much of the Detroit River shoreline is already hardened. However, there is growing interest in using soft engineering in appropriate locations. Soft engineering uses ecological principles and practices to achieve shoreline stabilization and safety while enhancing habitat, improving aesthetics, and saving money.

Hard engineering typically has no habitat value for fish or wildlife while soft engineering does. As noted in chapter 11, the Detroit River is one of the most biologically diverse areas in the Great Lakes Basin. Most local people who appreciate the outdoors know that the Detroit River supports a nationally renowned sport fishery. For example, the city of Trenton hosts a major walleye fishing tournament called Walleye Week, which attracts people from all over North America to compete in the In-Fisherman Professional Walleye Tournament, the Team Walleye Tournament, and the Michigan Walleye Tournament (over $240,000 in prize money is available). It is estimated that walleye fishing alone brings in $1 million to the economy of communities along the lower Detroit River each spring.

There are other reasons why soft engineering practices are being encouraged. It is well recognized that there is limited public access to the Detroit River, particularly on the U.S. side. Use of multiple-objective soft engineering of shorelines will increase public access to the river. There are also economic benefits. In general, soft engineering of shorelines is typically less expensive than hard engineering. Additionally, long-term maintenance costs of soft engineering are generally lower because soft engineering uses living structures that tend to mature and stabilize with time.

We have learned that it is important to redevelop and redesign our shorelines for multiple objectives. Shorelines can be stabilized and achieve safety while increasing public access, enhancing habitat, improving aesthetics, and saving money. Hard engineering of shorelines can cost as much as $1,200 or more per linear foot. We cannot afford to use hard engineering along the entire length of the Detroit River shoreline, nor do we want fully hard engineered shorelines because they have no habitat value and will not support the diversity of fish and wildlife that our river has blessed us with.

We have also learned that hard and soft engineering are not mutually exclusive; there are places where attributes of hard and soft engineering can be used together. This makes sense in a high-flow river like the Detroit River, through which the entire upper Great Lakes pass.

It is critically important that the right people get involved up-front in redevelopment projects to be able to incorporate principles of soft engineering into future waterfront designs. The design process must identify opportunities and establish partnerships early in the process that achieve integrated ecological, economic, and societal objectives. It is the hope of the Greater Detroit AHR Initiative that the advantages of soft engineering practices be recognized and incorporated into many shoreline projects along the Detroit River as the development standard into the future. Soft engineering demonstration projects have already been implemented at ten locations along the Detroit River (figure 14.3). Maheras Park in Detroit is one good example. A unique destination is being created with soft engineering along the shoreline of a new embayment, interpretive trails, and fishing piers (figure 14.4). The time is right to incorporate soft engineering practices into our efforts to redevelop and improve our shorelines and into municipal operating manuals and day-to-day operations. In addition, soft engineering projects lend themselves to volunteers. We need to work together to showcase the use of multiple-objective soft engineering practices along the shoreline of the Detroit River and proudly display our region's new front door.

It should also be noted that with the leadership of Congressman John Dingell, the Detroit River was designated as the first International Wildlife Refuge in North America in 2001. Despite the loss of over 95 percent of the coastal wetland habitats in the river, it still supports 65 species of fish and 29 species of waterfowl. The Detroit Audubon Society has identified 300 species of birds in the Detroit-Windsor area, with over 150 species nesting near the river. This designation as an International Wildlife Refuge helps raise the profile of the river's unique biodiversity and helps catalyze actions to conserve and rehabilitate key habitats that provide the foundation for our economies and sustain our quality of life.

Furthering Linked Riverfront Greenways

Urban infrastructure is generally understood to mean the substructure (e.g., roads, sewers) and underlying foundation that provide essential community services. Recently the concept of "green infrastructure" has been used to communicate the importance of the natural resource foundation that provides essential ecological services and related social and economic benefits.

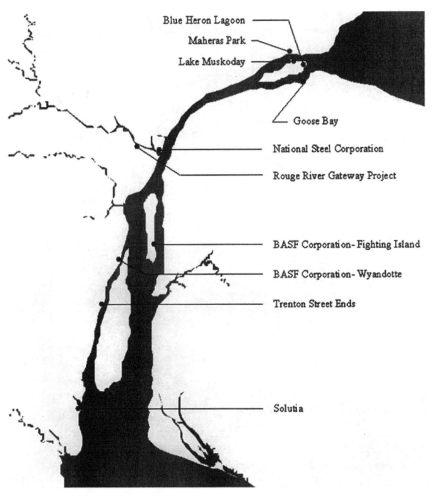

Blue Heron Lagoon
Maheras Park
Lake Muskoday
Goose Bay
National Steel Corporation
Rouge River Gateway Project
BASF Corporation- Fighting Island
BASF Corporation- Wyandotte
Trenton Street Ends
Solutia

Figure 14.3. Soft engineering demonstration projects being implemented along the shoreline of the Detroit River.

Parks, conservation areas, ecological corridors, linked greenways, and open spaces under best management practices all provide essential ecological services and related social and economic benefits. There are numerous examples around North America that demonstrate that putting money into green infrastructure is not merely a matter of paying for past mistakes but a sound investment with immediate and long-term returns. For example, through the work of the Toronto Waterfront Remedial Action Plan (RAP) and the Toronto Waterfront Regeneration Trust, full costs and benefits of restoration

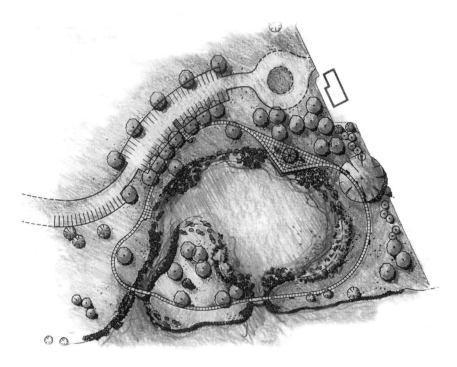

Figure 14.4. A diagram showing the improvements to the shoreline at Detroit's Maheras Park (Hazen and Sawyer).

projects in the Lower Don River Valley have been estimated (Muir, personal communication). Capital expenditure on watershed restoration is estimated to be about $964 million, along with $1.4 million in annual operating costs. (Figures here and below are in Canadian dollars.)

This will lead to capital savings of approximately $42 million, annual user benefits of approximately $55 million, and annual savings of approximately $11 million. In addition, the direct economic development benefits include province-wide increases in income of about $3.6 billion associated with capital investment and $5 billion per year associated with expanded economic activity. Such estimates of full costs and benefits help provide a compelling rationale for investing in green infrastructure.

There is growing recognition of the importance of green infrastructure in sustaining our communities, environments, and economies. Healthy communities and economies require healthy environments and the requisite green infrastructure. It is a key element in achieving community renewal,

Figure 14.5. General Motors' Wintergarden at the Renaissance Center facing the Detroit River and featuring greenways (Hines Development and Skidmore Owings & Merrill, LLP Master Architects).

increasing community awareness, participation, and pride, and sustaining communities and economies.

In southeast Michigan, green infrastructure is being created in the form of linked greenways. Embraced by communities throughout the region as the public place of choice, greenways have redefined what a quality community resource is all about (figure 14.5). No other resource serves a community in so many tangible ways. It is no wonder that the identity of many communities is tied closely to the quality of their waterfronts and greenways.

Greenways are corridors of linear open space established for recreation and/or conservation. They follow natural and cultural corridors, such as rivers, streams, scenic roads, and railroad corridors. Greenways may be held on public land, voluntarily retained on private land, or conserved through public-private partnerships.

It should be no surprise that stakeholders in the Greater Detroit AHR Initiative have identified linked greenways as one of their top six priorities. Again, people want to increase access to the Detroit River, incorporate trails and walkways to it, improve the aesthetic appearance of the shoreline, and reap recreational, ecological, and economic benefits from it. Our Detroit River has been rediscovered as an incredible asset and a key ingredient in achieving quality of life.

The industrial heritage of the area presents particular challenges and opportunities in establishing greenways. Historically, communities have been physically and emotionally separated from the river by a nearly continuous wall of commercial activity. The challenges of brownfield redevelopment are offset by the realization that one of the country's greatest resources can once again be part of the community. The idea of a nearly continuous greenway along the Detroit River seemed pure fantasy just a few years ago. Community and business leaders are now championing linked greenways along the entire length of the river from Lake St. Clair to Lake Erie, up key tributaries like the Rouge, Ecorse, and Huron Rivers, and across to Canada.

Given the choice, most people who live in cities would probably prefer to live near parks or greenways because of the obvious recreational, aesthetic, social, and ecological benefits. However, recent economic studies undertaken on Windsor's greenways have shown that property values generally increase with proximity to greenways (Zegarac and Muir 1996). Higher property values can also mean increased tax revenues.

Fantasy is now becoming reality along the Detroit River due to the coordinated efforts of dedicated individuals, businesses, and communities. A total of eighteen greenway projects have either been initiated or completed during 1999–2002 (see figure 14.6 and table 14.1). It is hoped that these linked greenway projects along the river will help catalyze additional ones.

These greenway projects are key linkages and unique destinations that provide open space, protect natural and cultural resources, improve the quality of life throughout our region, and serve as a unique legacy to future generations. They are also important because they bring people down to the shoreline and give them firsthand experiences with the river, which are essential to building support for further brownfield redevelopment, contaminated sediment remediation, pollution prevention, and habitat rehabilitation and conservation. These greenways also provide for alternative modes of transportation that further sustainable communities.

Concluding Remarks

Although only six first-phase priorities were set by the Greater Detroit AHR Initiative, it was agreed that no projects would be dropped from the original list of over 150 potential projects. Additional projects are addressed in the spirit of continuous improvement as time, interest, and resources become available (figure 14.1). However, it is important to set priorities on where to direct initial energies and limited resources.

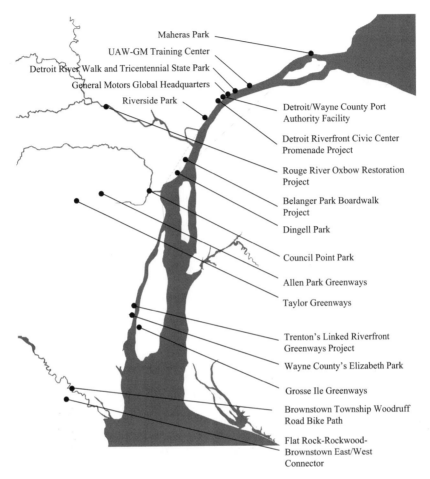

Figure 14.6. A map showing greenways developments along the Detroit River.

This book was written to help celebrate the three hundredth anniversary of European establishment of a settlement in the Detroit-Windsor area in 1701 (Detroit 300), the 2001 designation of the Detroit River as a Canadian Heritage River—making it the first international heritage river system in the world—and the 2001 designation of the lower Detroit River as the first International Wildlife Refuge in North America.

Indeed, the Detroit-Windsor metropolitan area is blessed with a rich history of culture, ethnic diversity, politics, industry, commerce, and ecology. Detroit 300 was a year-long, binational celebration featuring spectacular events, festivals, projects, and educational programs. Highlights included a

TABLE 14.1
**Descriptive and Cost Information of 18 Greenways Projects
along the Detroit River Shoreline**

Project: Maheras Park
Location: Detroit
Cost: $2.3 million
Description: The Detroit Water and Sewerage Department is undertaking a fish habitat mitigation project in this park during 2002–2004. Upon completion, this project will create a unique destination that will draw people to the Detroit River, improving fishing and other recreational opportunities where few recreational activities exist. It will enhance public access, rehabilitate habitat, and promote environmental education through interpretive signage.

Project: UAW-GM Training Center
Location: Detroit
Cost: Over $50 million
Description: The United Auto Workers and General Motors have created an exciting riverfront development that enhances Detroit River greenways. Visible from the Detroit River, this complex has a seven-story office tower with an adjacent three-story training tower and a 400–seat auditorium. The nine hundred feet of riverfront greenway provides public access and is a key element of the planned river walk to Belle Isle.

Project: Detroit River Walk and Tricentennial State Park
Location: Detroit
Cost: $50 million for the River Walk and $50 million for the state park
Description: On December 12, 2002, General Motors Corporation, the city of Detroit, and numerous partners, including the Greater Detroit AHR Initiative, announced an East Detroit Riverfront Vision that will construct a Detroit River Walk from Joe Louis Arena to the Belle Isle Bridge. The first state park in Detroit will be constructed on the east riverfront. It will create a "Gateway" experience to discover Michigan's State Parks, interpret natural resources and Michigan's rich economic history, connect Detroit's riverfront parks and adjacent neighborhoods, provide recreational opportunities for urban residents and visitors, and empower Detroiters to embrace conservation careers.

Project: General Motors Global Headquarters
Location: Detroit
Cost: $500 million
Description: This redevelopment project includes a Wintergarden flanked by two stories of retail space. Retail shops will line the river promenade and outdoor restaurant seating will create a new ambiance.

Project: Detroit/Wayne County Port Authority Facility
Location: Detroit
Cost: $7.5 million
Description: The Port Authority is building a multi-use public dock and passenger terminal facility located near Hart Plaza. This will be a unique destination and provide pedestrian access along Detroit's riverfront.

Project: Detroit Riverfront Civic Center Promenade Project
Location: Detroit
Cost: $6.2 million
Description: This project created approximately 3,500 lineal feet of city-owned riverfront greenway in the Civic Center area. This promenade provides an aesthetically pleasing pedestrian link between the newly renovated General Motors Global Headquarters, complete with new hotel, retail shops, and restaurants, and the Cobo Conference/Exhibition Center.

Project: Riverside Park
Location: Detroit
Cost: $1 million
Description: Detroit Recreation Department repaired the seawall and riverfront promenade beginning in 2001. This park offers one of the best places to watch freighters.

Project: Rouge River Oxbow Restoration Project
Location: Dearborn
Cost: $2 million (first phase of oxbow restoration)
Description: The Rouge River Gateway Partnership is restoring a historical oxbow; relocating a combined sewer overflow discharge pipe that impacts the oxbow; and establishing a greenway trail system that will connect the developed portions of Greenfield Village to the wetlands, oxbow, and channel portion of the Rouge River.

Project: Belanger Park
Location: River Rouge
Cost: $1 million
Description: A two-phase restoration of a 9.5–acre park was completed in 1999 and now offers improved access for fishing, boating, picnicking, and special celebrations like family reunions.

Project: Dingell Park
Location: Ecorse
Cost: $500,000
Description: This project will improve the boardwalk and connections to the adjacent neighborhood.

Project: Council Point Park:
Location: Lincoln Park
Cost: $500,000
Description: Another one mile of trail, an inline skating arena, and a comfort station were completed in 2000 at this park located at the confluence of the three branches of the Ecorse Creek.

Project: Allen Park Greenways
Location: Allen Park
Cost: $300,000
Description: This project will create trails that will pass through and connect most of the city's major parks.

CHAPTER 14

Project: Taylor Greenways
Location: Taylor
Cost: $1.5 million
Description: This greenway trail beautification project will run along Goddard Road from Telegraph to Allen Road.

Project: Trenton's Linked Riverfront Greenways Project
Location: Trenton
Cost: $675,700
Description: Greenway connections will be established beginning in 2001 among the existing parks, three new pocket parks at street ends along the Detroit River, and downtown in an effort to further economic development and enhance community pride.

Project: Wayne County's Elizabeth Park
Location: Trenton
Cost: $1 million
Description: This 2003 project in Michigan's oldest county park stabilized 600 feet of shoreline; created a new riverwalk with seating, lighting, handicap access, and fishing sites; and enhanced fish habitat.

Project: Grosse Ile Greenways
Location: Grosse Ile
Cost: Over $40,000 per year since 1994
Description: Through dedicated millages, Grosse Ile Township has created approximately 8.6 miles of bike path since 1994 and added an additional 2.5 miles in 2001. Grosse Ile Township has also invested over $450,000 per year on preserving open space since 1997.

Project: Brownstown Township Woodruff Road Bike Path
Location: Brownstown Township
Cost: $340,000
Description: A new 1.5 mile bike path was constructed in 2003 to help link several communities with Huron Clinton Metropolitan Authority Parks.

Project: Flat Rock-Rockwood-Brownstown East/West Connector
Location: Flat Rock, Rockwood, and Brownstown Township
Cost: $2.2 million
Description: A five-mile greenway trail will connect 4,800 acres of parkland in the Huron River watershed with the lower Detroit River and 4,000 acres of state game land.

reenactment of Cadillac's landing in downtown Detroit; the tall ships parade; a $15 million Legacy Project that beautified Detroit's neighborhoods and civic core (planting trees in neighborhoods and parks; creating school-yard habitats; enhancing Woodward Avenue from Hart Plaza on the Detroit River up to Grand Circus Park; adding new art, historical markers about Detroit's maritime history, a monument to the Underground Railroad, a statue of Antoine de la Mothe Cadillac, and other amenities to Detroit's Riverfront Promenade; and creating a conservancy endowment to ensure the high-quality maintenance of downtown and neighborhood project improvements); a Canada–United States State of the Strait conference (*http://cro nus.uwindsor.ca/greatlakes*); and the formal designation of the Detroit River as a Canadian Heritage River.

The Detroit River and its waterfront were center stage throughout Detroit 300. However, to truly leave a legacy for future generations we must continue to use the community-based models of the Greater Detroit AHR Initiative and the Canadian Heritage River Initiative (Essex Region Conservation Authority 1999) to further balanced and continuous economic, environmental, and community progress. Both of these heritage river initiatives have engendered considerable support and are working to coordinate efforts to achieve our common goals.

To ensure these community-based models remain successful, we must all develop a respect for the place or ecosystem we call home. Aldo Leopold, in his pioneering work on an American land or stewardship ethic, articulated best this need to respect and honor the place we call home:

> All ethics so far evolved rest upon a single premise: that the individual is a member of a community of interdependent parts. . . . The land ethic simply enlarges the boundaries of the community to include soils, waters, plants, and animals, or collectively: the land. . . . In short, a land ethic changes the role of *Homo sapiens* from conqueror of the land-community to plain member and citizen of it. It implies respect for his fellow-members, and also respect for the community as such. (1949, 239–40)

In addition, we must become caretakers of the place or ecosystem we call home. There is a fundamental difference between environment and ecosystem. Environment is the physical and chemical setting in which life exists. It is often thought of something separate from animals and plants. Life, including humans, is considered part of an ecosystem. It is like the difference between house and home. House is something external and detached.

Home is something we see ourselves in even when not there. We must come to see ourselves as part of our Detroit River ecosystem and we must learn to care for it as our home. It can start with adopting and caring for our own backyard or favorite park and can grow to include whole communities and river ecosystems.

As part of our caretaker responsibility, we must also learn to think intergenerationally like the Iroquois, who believe that for any action people need to consider the effects on the seventh generation to follow. Oren Lyons of the Onondaga Nation of the Haudenosaunee (Iroquois) shares this timeless message of his people: "Take care how you place your moccasins upon the earth. Step with care, for the faces of the future generations are looking up from the earth, waiting their turn for life" (Palmer 1999, 9).

Our hope is that this book has honored the Detroit River as one of the world's greatest rivers, educated people on the history of our ecosystem problems associated with the river, documented progress in addressing our ecosystem problems, identified key factors and efforts required to protect and rehabilitate the river, and elicited personal commitments to honor our Detroit River and care for our home.

It will only be through a change in attitude toward the river that we will be able to achieve our individual and collective goals and leave a true legacy for future generations. Again, Golding (1998) has stated it best: "To deny the river is to deny the origin of the city. To rethink the river is to discover a unique opportunity to define urban places, join neighborhoods and communities together, and reconnect us to our landscape and our history" (5).

LITERATURE CITED

Carson, R. 1962. *Silent Spring.* Boston: Houghton Mifflin.

Caulk, A. D., J. E. Gannon, J. R. Shaw, and J. H. Hartig. 2000. *Best Management Practices for Soft Engineering of Shorelines.* Detroit: Greater Detroit American Heritage River Initiative.

Essex Region Conservation Authority. 1999. *Detroit River Nomination Document: Canadian Heritage Rivers System.* Essex, Ont.

Global Environment & Technology Foundation. 1999. *Proceedings of the National Town Meeting on a Sustainable America.* Annandale, VA. *http://www.sustainableusa .com.*

Golding, A. 1998. *The Los Angeles River: Reshaping Our Urban Landscape.* Los Angeles: Target Science.

Leopold, A. 1949. *A Sand County Almanac.* New York: Oxford University Press.

Muir, T. *Environment Canada, Canada Centre for Inland Waters.* Burlington, Ont.

Palmer, T. 1999. *The Heart of America: Our Landscape, Our Future.* Washington, DC: Island Press.

Reid, R., K. Rodriguez, and A. Mysz. 1999. *State of the Lakes Ecosystem Conference 1998: Biodiversity Investment Areas—Nearshore Terrestrial Ecosystems.* Chicago: U.S. Environmental Protection Agency; Burlington, Ont.: Environment Canada.

U.S. Environmental Protection Agency (U.S. EPA) 1999. *EPA's Framework for Community-Based Environmental Protection.* EPA 237-S-99–001. Washington, DC: Office of Policy.

Zegarac, M., and T. Muir. 1996. *A Catalogue of Benefits Associated with Greenspaces and an Analysis of the Effects of Greenspaces on Residential Property Values: A Windsor Case Study.* Burlington, Ont.: Environment Canada.

CONTRIBUTORS

BEAN, CATHERINE J.   Cathy graduated in 1986 from the University of Michigan-Dearborn with a bachelor's degree in environmental science. She has worked for the Michigan Department of Environmental Quality for fifteen years in the Surface Water Quality Division. Her responsibilities have included monitoring compliance of wastewater dischargers with state and federal regulations, as well as heading up the implementation of the Rouge River Remedial Action Plan. More recently she developed and implemented a one-of-a-kind voluntary storm-water permit for the Michigan Department of Environmental Quality for communities and agencies in the Rouge River watershed.

BULL, JAMES N.   Jim, a private environmental consultant, holds a Ph.D. in natural resources from the University of Michigan. He is past president and current board member of the Detroit Audubon Society and board member of Friends of the Detroit River. He has helped coordinate educational programs for the Huron, Rouge, Cuyahoga, Saginaw, and Clinton Rivers, and is a cofounder of the Global Rivers Environmental Education Network.

CIBOROWSKI, JAN J. H.   Jan is a professor in the Department of Biological Sciences and research scientist in the Great Lakes Institute for Environmental Research, University of Windsor. He received his graduate training at the Universities of Toronto and Alberta, studying the ecology and behavior of stream insects. He presently investigates the influences of environmental stresses on development, distribution, and community composition of aquatic invertebrates. He is a founder and co-director of the Lake Erie Millennium Plan research network.

CORNELL, GEORGE L.   George is a professor of history and American studies, as well as the director of the Native American Institute, Center for

Urban Affairs at Michigan State University. He is a specialist in Native American history and contemporary affairs, with extensive experience in public service to Indian organizations. His current interests include expanding university services to Native American communities, conducting research on the American Indian as a conservationist, and studying the interpretation of Indian oral traditions.

CRAVES, JULIE   Julie is the supervisor of avian research at the University of Michigan-Dearborn's Rouge River Bird Observatory. She is also the editor of the state journal *Michigan Birds and Natural History,* a contributing editor to *Birder's World* magazine, and coauthor of the updated breeding bird atlas for Michigan.

DOLAN, DAVID M.   Dave teaches undergraduate and graduate statistics courses in the Department of Natural and Applied Sciences at the University of Wisconsin-Green Bay. Dave earned his doctorate in mathematics at McMaster University in Hamilton, Ontario, in 1999. He has a masters degree in statistics from the University of Michigan (1980) and in environmental engineering from the University of Notre Dame (1972). His major scientific interests are the application of statistical techniques to water quality problems in trend detection (temporal and spatial) and the evaluation of the effectiveness of pollutant reduction efforts. He currently is an associate editor for the *Journal of Great Lakes Research* and chairman of the Nominations Committee of the Board of Directors for the International Association for Great Lakes Research.

GIVENS-McGOWAN, KAY   Dr. McGowan is a cultural anthropologist. She has served as the Federal Recognition Project Director of the Wyandot of Anderdon Nation. Kay is of Cherokee/Choctaw ancestry. She teaches Native American studies at Marygrove College where she is an assistant professor, and she is also an adjunct professor at Eastern Michigan University.

HAFFNER, DOUG   Doug completed a Ph.D. at the University of London, England, and upon returning to Canada spent the next twenty-five years studying the Great Lakes. As the former director of the Great Lakes Institute for Environmental Research at the University of Windsor and a professor of biological sciences, he has published numerous scientific articles on contaminant distributions and dynamics in the Detroit River. He currently is the chair of the Contaminated Sediments Subcommittee of the Detroit River

Canadian Clean-up Committee. Recently he has been appointed to the Senior Canada Research Chair in Great Lakes Environmental Health.

HARTIG, JOHN H.   John obtained his Ph.D. in limnology from the University of Windsor and has over twenty years of practical experience in environmental management. He currently serves as river navigator for the Greater Detroit American Heritage River Initiative, where he works with communities and businesses to identify and implement high-priority projects related to environmental stewardship, economic development, and celebration of history and culture. He also is an adjunct professor at Wayne State University where he teaches environmental management and sustainable development. He is also a past president of the International Association for Great Lakes Research.

KERR, JOHN K.   John has worked for the Detroit/Wayne County Port Authority as an economic development specialist since 1998. He has helped the Port Authority secure $7.5 million in federal and state appropriations for a public dock and passenger terminal on the waterfront in Detroit. This state-of-the-art facility will be able to receive cruise ships and other transient vessels. John is a graduate of Michigan State University, holding a bachelor of science degree in urban and regional planning. He was involved in internships with the City of East Lansing Planning and Community Development Department and for Second Ebenezer Baptist Church of Detroit where he worked on faith-based partnerships with Comerica Bank. John is an active member of the American Planning Association and the Michigan Economic Developers Association.

MANNY, BRUCE A.   Dr. Manny has worked as a research fishery biologist for the U.S. Geological Survey and the U.S. Fish and Wildlife Service for the past twenty-eight years. Much of his research has been on fish, wildlife, and their habitats in the Detroit River. His current research interest is restoration of habitat for native fish and wildlife in the Detroit River, including lake sturgeon and diving ducks.

MULLETT, NOEL, JR.   Noel is the Rouge Project Technical Coordinator for Wayne County's Department of Environment in the Watershed Management Division. In this role, he assists the director with the coordination and implementation of the Rouge River National Wet Weather Demonstration

Project and with development and implementation of watershed management services including the General Stormwater Permit compliance, illicit discharge elimination, public education/pollution prevention, watershed planning, monitoring, and restoration.

MURRAY, PAT   Pat is a freelance writer who has lived for a decade on each side of the Detroit River. Recently, she was a key organizer of the 2001 State of the Strait Conference at the University of Windsor that focused on evaluating the status and trends of key indicators for the Detroit River ecosystem.

OLINEK, W. STEVEN   Steve is a twenty-five-year veteran of the Detroit shipping industry and currently is the deputy director of the Detroit/Wayne County Port Authority. In this position, he is responsible for, among other things, the marketing and administration of Greater Detroit Foreign Trade Zone, Inc., one of the largest and most successful foreign-trade zone programs in the country. Since beginning his career with the Grand Trunk Western Railroad, Steve has worked in virtually all facets of the trade-development and intermodal transportation industries, from trucking to freight brokerage to corporate logistics with the Stroh Brewery Company, where he also served as director of development for the company's private fleet.

PANEK, JENNIFER   Jennifer recently completed her bachelor's degree in environmental studies at the University of Michigan-Dearborn. While attending the University of Michigan-Dearborn, she worked in the Natural Areas Department where she led educational programs that focused on natural resources and the environment. While at University of Michigan-Dearborn she also did an internship with the Greater Detroit American Heritage River Initiative where she worked on Detroit's role in reversing cultural eutrophication in Lake Erie. Jennifer has also served as a provisional interpreter at Lake Erie Metropark.

READ, JENNIFER   Jennifer is currently assistant director of Michigan Sea Grant. Prior to this position she served as a research associate at the Great Lakes Institute for Environmental Research, University of Windsor. She is a specialist in Great Lakes pollution policy and, among other issues, works on the Detroit River Remedial Action Plan.

STAFFORD, TERRY   Terry currently is a student at Henry Ford Community College where he is specializing in environmental and natural sciences. As part of his academic training he worked on an American Heritage River internship researching historical oil pollution problems in the Detroit river. Terry is currently employed by the state of Michigan's Social Security Administration.

STEINMETZ, D. C.   Dan has lived and worked in the Detroit area his entire life. He received a bachelor of science degree in biology from Wayne State University and a master of science degree in occupational and environmental health from the Wayne State University School of Medicine. He manages the Product Stewardship program of a large chemical company, where he has spent more than twenty years in the ecology and safety areas.

THIEL, D. P.   Doug is a lifelong downriver resident. He has had a twenty-five-year career at a major chemical company in the Detroit area and currently serves as an environmental, health, and safety manager covering the North American Free Trade Agreement region. In addition, he has worked as an adjunct professor at Oakland University for twenty years teaching a variety of industrial hygiene and emergency response courses. Doug holds a bachelor of science degree from Central Michigan in Biology and a master of science degree from Wayne State University in occupational and environmental health. He is a board member with the Grosse Ile Nature and Land Conservancy and is an active member of the Friends of the Detroit River.

INDEX

Aatisken, 26
Aboriginal people: Algonquins, 9–11, 15; Erie Nation, 30; Iroquois, 10, 11, 18, 222; Potowatomi, 28; Tecumseh, 31; Wyandot, 23–27, 32, 121
Aki, 26
Algonquins. *See* Aboriginal people
Ambassador Bridge, 45, 46
Amphipods, 109
Antiquites Act, 200
Aphanizomenon flos-aquae, 87
Aquatic environment, 150
Asken, 26
Atieeronnon, 24
Atlantic and Mississippi flyways, 70
Atlatl, 12
Atsihiendo, 28

Baraga, Father, 10
Battle of Monguagon, 31. *See also* Wyandot
Beaver, 14, 23, 49, 50–58, 155, 199. *See also* fur trade
Belle Isle, 7, 64, 144, 147–49, 155, 159, 161, 165, 166, 205, 206
Benthic invertebrates, 107
Biodiversity, 142, 144, 145, 161, 162, 164, 212
Biodiversity Investment Area, 6
Bivalves, 108
Black Lagoon, 209
Brownfields, 208
Burbot, 28

C3 (Clean Corporate Citizen), 179
Caddisflies, 109, 116
Cadillac, Antoine de la Mothe, 18–20, 23, 36, 55, 59, 124, 142, 146, 147, 157, 221
Canadian Heritage River Initiative, 4, 217, 221. *See also* Greater Detroit AHR
Canard River Marsh, 71, 149
Carrying capacity, 54, 55
Carson, Rachel, 91, 174, 200
Cartier, Jacques, 41
Cheumatopsyche, 109
Chironomus, 116
Cholera, 60–62
Cladophora, 79
Clean Water Act, 172–74
Colonial jellyfish, 107
Common Tern, 143, 144, 160, 164
Contaminated sediment, 209, 216
Cordylophora caspia, 107, 110
Creator, 15, 18, 21
Cryptosporidium, 67
CSO (Combined sewer overflow), 66, 67, 75, 185, 189–93
Cultural eutrophication, 79, 84, 88
Curly-leaf pondweed, 151

Daphnia lumholtzi, 152
DDT, 91, 93, 200
Depostional zones, 110
Design for Environment, 175, 179
Detroit Audubon Society, 142, 212
Detroit River RAP (Remedial Action Plan), 127, 130

Detroit Wastewater Treatment Plant, 6, 81–85
Discharge rate, 6
Drinking water, 60, 62, 63, 67, 172
Ducks, 70–73, 78, 156–59
DWSD (Detroit Water and Sewage Department), 66

Echinogammarus ischnus, 107
Eco-efficiency, 175, 182
Ecogifts, 132
Ecosystem, 1–3, 58, 199, 210, 221, 222
Ekarenniondi, 26
Elizabeth Park, 148
Environment, 2, 221
Environment Canada, 124
Erie Nation. See Aboriginal people
Erosional zones, 108
Eurasian Watermilfoil, 150–51
Eutrophic status, 79
Eutrophication, 79–81, 83, 172

Feast of the Dead, The, 26. See also Aatisken; Aboriginal People: Wyandot; Aki; Asken
Ford, Henry, 69, 176, 181, 182, 186
Forests, 145, 147
Fort Wayne, 7, 207
Friends of the Detroit River, 32
Friends of the Rouge, 189, 197
Fur trade, 11, 18, 23, 30, 36, 53, 55–57, 122, 155. See also beaver

Gastroenteritis, 64
Grassy Island, 160, 162
Great Lakes Fishery Commission, 89
Great Lakes Water Quality Agreement, 81
Greater Detroit AHR (American Heritage River) Initiative, 4, 202, 203, 205, 207–12, 215, 216, 221
Green chemistry, 183
Green Infrastructure, 212, 214, 215
Griffin, 35, 36
Gulls, 160

Habitat, 124, 126, 127, 130–36, 146, 212
Hard engineering, 211

Hennipen, Father, 18, 35, 124
Hexagenia, 87, 111, 114, 116, 117
Humbug Complex, 147, 149, 154, 163, 165
Humbug Marsh, 142, 147, 149, 150, 161, 163
Hydropsyche, 109
Hydrozoans, 110

IJC (International Joint Commission), 7, 42, 62, 64, 65, 100, 127, 135
International Wildlife Refuge, 132, 212, 217
Invertebrates, 151
Iroquois. See Aboriginal people

Johnson Creek Protection Group, 193

La Salle, 35, 36
Lake Erie, 10, 19, 26, 27, 35, 37, 41, 71, 79–81, 83–84, 86–89, 96, 97, 112, 113, 122–24, 154, 156, 166, 171, 216
Land birds, 159, 161
Leinchataon, 28
Leptospirosis, 188
Longhouses, 26, 28

Macroinvertebrates, 187
Macronema zebratum, 110
Marsh birds, 160
Mayflies, 6, 87, 116–17
MBP3 (Michigan Business Pollution Prevention Partnership), 177
MDEQ (Michigan Department of Environmental Quality), 173, 177, 179, 187, 189, 190, 193–95
Mercury, 91–102
Mesotrophic status, 79
Methyl mercury, 94–96
Michigan Department of Community Health, 6
Michigan Department of Natural Resources, 74, 76, 96, 101, 135, 142, 187, 188, 195, 210
Microcystis, 88
Mother Earth, 16
Muskrats, 12

Narrows, 55
Native Americans, 3, 49, 51, 53, 55–57,
 121, 122. *See also* Aboriginal people
Native people, 9–21, 27, 186. *See also*
 Aboriginal people
New France, 10, 35
Nitellopsis obtusa, 151
Nonpoint source, 189, 191
North American Waterfowl Management
 Plan, 124
NPDES (National Pollutant Discharge
 Elimination System), 174, 187
Nuisance algal conditions, 89
Numma Sepee, 24

Oil pollution, 70, 72, 74–76, 78, 199
Oil spill, 77, 78
Oki, 27
Oppenago, 24
Over-exploitation, 49, 51, 53, 57

PCB (Polychlorinated biphenyls), 91, 93,
 94, 98–102
Phosphorus, 80–84, 87–89, 199
Pollution tolerant, 185, 187
Port of Detroit, 40, 41
Potowatomi. *See* Aboriginal people
Prairies, 145, 146

Quagga, 151, 152
Quagga mussels, 88

Raptors, 156, 157
Raw sewage, 66, 185, 187
Removal Period, 32
Responsible Care®, 176
Robiche, 24
Roosevelt, Theodore, 200
Rouge Gateway Partnership, 195
Rouge River Basin Committee, 188, 189
Rouge River National Wet Weather Demon-
 stration Project, 189, 190
Rouge River RAP (Remedial Action Plan),
 188–90
Rumrunners, 46
Rusty crayfish, 152

Sagamite, 28
Sagard, Father, 29, 30
Saint Lawrence Seaway Development Cor-
 poration, 42
Salmonella typhi, 63
Sanitary sewers, 185, 189–91
Seaway idea, 42
Sediment Remediation, 102, 103, 210
Shipbuilding, 3, 6, 43, 69, 172, 186
Silent Spring (Rachel Carson), 91, 174, 200
Soft engineering, 210–13
Spiny amphipod, 107
Spiny Waterflea, 152
"Spirit of Trenton," 135
Standpipe tower, 62, 63
Storm water, 66, 189, 193
Sturgeon, 13
Suckers, 13
Sustainable development, 175, 179, 184
Swan Island, 148

Tahounehawietie, 27
Tecumseh. *See* Aboriginal people
Teuchsay Grondie, 29
Toh Roon Toh, 26, 27
Tonquish, 24
Toxic chemicals, 92, 100–102
Toxic substances, 122, 131, 199
Trade, cross-border, 45–47
Tragedy of the Commons, 54, 55
Trail of Tears, 32. *See also* Aboriginal people
Tsugsagrondie, 29
Typhoid epidemic, 172
Typhoid fever, 63–66

U.S. Army Corps of Engineers, 127, 134,
 163, 207, 210
U.S. EPA (Environmental Protection
 Agency), 124, 127, 135, 164, 172, 173,
 209, 210
U.S. Geological Survey's Great Lakes Science
 Center, 127
Upper Great Lakes Connecting Channel
 Study, 100
Use impairment, 128, 129, 136

Voyage, 18, 19
Voyageurs, 186

Walleye, 87, 88, 93, 96
Wastewater, 66
Waterborne disease, 60, 62, 64, 65, 69, 199
Waterfowl, 11, 12, 17, 70–72, 76, 77, 142,
    157–59, 212
Watershed, 1, 2, 49, 55, 57, 185–92
Waterworks Park, 62
Wawiiatanong, 9–21
Wetlands, 2, 12, 123–27, 130, 131, 133,
    134, 142, 148, 149

Whitefish, 39
Wild celery, 158
Windsor-Detroit Tunnel, 45, 46
Winter duck kill, 72, 74, 76, 78, 172
WWTP (Wastewater Treatment Plant), 64,
    66, 67, 75
Wyandot. *See* Aboriginal people
Wyandotte National Wildlife Refuge, 71

Yandatssa, 25

Zebra mussels, 88, 151, 152, 162
Zoobenthos, 107, 108
Zooplankton species, 88